JUNGLE JUICE

Portrait of a Career – in Oils

Although this narrative is biographical to some degree,
all characters portrayed are fictitious.
Any resemblance to real people would be entirely coincidental.

Eric Goodger

Dedicated to all members of my family,
past and present,
and to Charlie Green of Sunbury.

Copyright © 2005 E Goodger

Produced in association with

Bound Biographies

Heyford Park House, Heyford Park
Bicester, Oxon OX25 5HD
www.boundbiographies.com

and

Landfall Press

egoodger@waitrose.com

ISBN: 0 95201869 1

Acknowledgements

Guy Poltock for the design of the cover illustrations and the cartoons.

Brian Hunt for the photograph of the author 'riding' a V2 rocket.

Image 5 for the photograph of the Re-enTree oak.

Mike Oke for continual wise guidance and occasional gentle bullying.

Introduction

This is the story of Matt Gregg, a London-born student of aeronautical engineering, who served in the RAF and then decided to specialise in fuel technology. He has to work hard to learn but perseveres with much determination, and succeeds as far as his talents permit.

He has a ready wit, and the elements of leadership which hold him in good stead for his career in the Forces and academia. His greatest asset is a deep resource of common sense, and he is also an effective teacher. He is guided by strong moral principles, and is emotionally dependent on marriage. His controversial views are firm but not fixed, and he is always ready to listen.

His life's journey through joys and sorrows, successes and failures, has moulded him into a compassionate citizen and friend.

Its recording may prove both entertaining and uplifting.

E G

> You call him 'engineer' in a taunting sense,
> and would refuse either to bestow your daughter on his son
> or let your son marry his daughter.

> PLATO

Contents

9	Introduction
	Chapter 1 Michael (1940)
27	Chapter 2 Matthew (1940)
37	Chapter 3 OTU (1940)
45	Chapter 4 Watershed (1941)
55	Chapter 5 Brickhill (1941)
71	Chapter 6 Kismet (1942)
83	Chapter 7 Bonded (1943)
91	Chapter 8 Postings (1943)
99	Chapter 9 Overlord (1944)
117	Chapter 10 EVT (1945 – 46)
131	Chapter 11 College (1947 – 50)
141	Chapter 12 Oil (1951 – 54)
153	Chapter 13 Canals (1954)
169	Chapter 14 University (1955 – 60)
183	Chapter 15 Oz (1960 – 65)
203	Chapter 16 Survey (1966 – 69)
217	Chapter 17 Return (1969 – 79)
229	Chapter 18 Loss (1979 – 89)
249	Chapter 19 Gain (1990 – 97)
279	Chapter 20 Viewpoints (1998 – 2000)
301	Chapter 21 Unknown (2000 – 04)
323	Chapter 22 Aftermath (2004 – 05)
339	Chapter 23 Phone (2005)
Appendices	
341	Appendix 1 Preparing for Science-based Examination Success
345	Appendix 2 Touch Typing
347	Appendix 3 Matt Gregg's Method of Teaching Fuel Technology
353	Appendix 4 The City of Bath (1988)
369	Bibliography
373	Index

Chapter 1 – Michael (1940)

MIKE didn't see the bandit that hit him. Having just lined up his sights on an ME109, with his thumb poised over the firing button, he was shocked when his instrument panel exploded in a shower of splinters, and his right leg suddenly went to sleep. It was high summer, with the Battle of Britain at its peak.

Christ, he thought, *I'm hit. Get the hell out of here.*
He banked as steeply as he dared, and searched desperately for some cloud cover.
No dice. Keep going – lose some height before everything else falls apart.

It was then that he noted in his mirror an ominous shape squaring up to finish him off. Just when preparing to eject, while he still might, he saw the shape erupt into a ball of fire and smoke. Seconds later an accompanying Spitfire streaked past, waggling its wings in recognition.
Good old Terry.
It was his wingman, keeping tabs on him.

Must stand him a beer when I get back – if I get back. But I've got to. Can't let the side down now. The kite still seems to fly. Responds to the stick. No sign of smoke. The old Merlin's still banging away, so I might make it. Let's see how many of the taps are still working. Throttle's a bit dodgy, and there's no revs or boost showing, and no radio, but at least the compass and the altimeter seem OK. That's something. The undercart lever's kaput so it's eject or belly flop. But this leg's starting to pain now. Hitting the deck hard on the end of a 'chute won't do it any good at all. Head north, old son, and try to make Good Old Bloody Old Blighty. Don't fancy wallowing in the drink waiting for Air Sea Rescue. Must get her back. Ginger Beer George told us he wanted all the aircraft bits he could get his mitts on – desperate shortage – must try and get down in one piece. What was it P/O Prune used to say? "A good landing is one you can walk away from," so if you must prang make it a good 'un. Bringing home the bits for you George. If only the tail wheel survives it'll be something.

Hello – starting to feel a bit dizzy. Loss of blood perhaps. Keep the oxygen going – if there's any left. Try and wrap your trousers tighter round that leg. One thing, this Spit has a metal prop – could help when we hit. Cripes, she starting to vibrate. Got to get her down. Still losing height nicely. Nothing on my tail, thank God. The boys are all mixing it up there over mid-Channel. Coastline coming up. Good job the Ack Ack characters know a Spitfire shape when they see it – probably better than I do myself. Shake your head and think clearly.

Countryside's coming up fast now. Can't hope for an airfield – ours or anybody's. Any flat piece of geography will be welcome right now. Ha – there's a field to starboard – looks flattish – no corn or anything to catch fire. Ye gods, yes. Hope there's no fire. With this leg I wouldn't have a prayer. No flaps so must dive – and engine running fast, but throttle still shaky – I'll just have to switch both mags off before she bumps. Slide canopy back and lock. Field coming up now. Steady, old girl. Here goes for a wheels-up. Try not to ground loop. Up comes the grass. Mags off. Wow.

GERDOING! GERDOING! GERDOING!

The stricken aircraft hit the ground, bounced several times then slid ingloriously along the wet grass to a steaming standstill. Petrol spilled from the wing tanks but, fortunately, there were no friction sparks so no fire.

Got to get out of here – fast.
Mike unbuckled, opened the door flap, and prepared to ease himself out of the cockpit. But it was going to be difficult with a useless leg.
What I need right now is a crane – or a rocket up my backside. Grit your teeth, and move it.

After the clamour and excitement of the last few minutes, the silence was like balm to the soul, but Mike was in no shape to enjoy it. Reaction began to set in, and he had a splitting headache. He ripped off his helmet and tried to draw in fresh breaths, but was engulfed with petrol fumes.
You're a big boy now so just get that bloody leg out of the cockpit.
He grabbed the windscreen and tried again to ease himself up.
Damn. It just won't work. I need a tin opener to get out of this.

"Hullo there! Hullo!" Mike was dimly aware of approaching voices, and clumping footsteps over wet grass. Suddenly a breathless, ruddy face appeared next to him, hotly pursued by a thick, gorgeously-shaped jersey of a land worker, undoubtedly of the feminine variety. "Are you OK? Can you move?" All in one breath.

"Well, sort of yes and no, I guess. If you can just help me out of this crate, I think I'll manage."

Mike then felt the gorgeous jersey jammed tight against his head as the land girl reached over to pull his legs up, but with little success and much pain.

"Hold on there." Another welcoming voice was heard, this time distinctly masculine, if a little aged. Very shortly, a second pair of hands appeared over the other side of the cockpit, and their combined efforts resulted in Mike being extracted like a bad tooth and laid on the grass. "For my next trick," muttered Mike, "I will dive off the roof into a wet sponge." Then, more urgently, "My thanks to one and all but could we get away from this hardware – in case she decides to blow up any minute?"

Together the farmer and his land girl dragged Mike well away, and then stopped to survey the situation. "There's a cottage hospital nearby," said the farmer, "and we'll get you there right away. Sorry I can't offer you the tractor – it's out of action, but I've got a horse and cart over the hedge. Look after him, Doris, while I go and get it."

And so Mike travelled the next leg of his journey in a relatively sedate fashion, changing down from nearly a thousand to, literally, one horsepower in the process. Not that he knew much about it. Waves of faintness washed over him, interspersed with bouts of consciousness and pain. *And what are all those coloured lights doing, flashing in front of me?*

* * * * *

Eventually Mike roused to find himself lying on some sort of couch with his trousers off, and an elderly gent in a white coat doing something to his leg. "Oh! There you are," said the doctor. "Looks like you're going to be OK. Flesh wounds mostly, but we're checking up on the bones. The RAF ambulance will be here shortly, but you'll be fine here till then. Try to relax now. Nurse will be along in a moment to see you – give you something to help you sleep."

"Thanks, Doc. What happened to my aircraft? Did it catch fire?"

"Don't think so. We can't see your landing spot from here, but there's no smoke in the sky."

"Thank God for that."

Maybe George will get something out of it after all.

Mike lay back and drew a deep breath. *Still that bloody headache – and the smell of petrol in my nostrils. Makes me want to vomit. But what else is it that I can smell? Ether? Surgical spirit? What was it that these medical characters splashed about?* He gazed with mild interest at the way the light reflected off the doctor's bald head. Then at the drab curtains surrounding the couch. *Might as well rest my eyes if that's all there is to see.* Mike did so with mixed feelings of relief and gratitude. He had done all he could for the moment. So now to sleep. *Jeez. Never felt so tired.*

Weeks of excitement, action, danger and drama had left him with a marked shortage of adrenaline, but sleep did not come easily. His brain still raced with the events of the day. Like an action replay, he was back in the cockpit – *Splintering glass. Aiming north. Vibration setting in. Must lose height. Where's a field? Any field? Over there. Coming up fast. Wait for it, wait for it. Screams of tearing metal. Must get out. Body trembling. Can't get a purchase. Helping hands. Lying on grass. Dragged away. Bumpy ride.*

Merciful sleep eventually took over as his body started the long process of recharging its batteries.

* * * * *

When he next roused, Mike was in a bed – in some sort of nightgown. Leg tingling rather than paining. The dressing seemed to give it comfort.

"Welcome back," came a cheerfully restrained voice. "Now I want you to sit up a bit and drink this." Mike was dimly aware of the presence of the crisp blouse of a nurse's uniform, and an arm supporting his shoulders. With a

mug placed to his lips, he did his best to sup and swallow. *God! What a foul taste. If that's the best the hospital cat can do, it ought to be put down.*
"Have you got something for a chaser?" he mumbled, with little hope.
"Now lie back and rest." The pleasant voice had the edge of authority in it now, so Mike followed orders as he was tucked back into bed. *Why fight it! That's enough for one day, anyway.*

* * * * *

Some time later – and Mike had lost all trace of time – he awoke to the presence of the nurse busily arranging things on his bedside locker. *My word, but that's a nice looking piece of skirt, isn't it? Damn it, she's more than that. Downright pretty I'd say.*

And so she was. Dark eyes, rounded pink cheeks and a mouth like a baby's. And all with dark hair curling from under her nurse's cap and, as far as Mike could ascertain from his less than vantage point, trim waist and a very neat figure to go with it all. Those ankles looked good too, especially in black stockings.

"There you are again," she murmured. "Doctor says you are to take this. It'll make you feel more comfortable and ready for the trip back to your base."

Mike sat up prepared to do as he was told. That way he could see more of her, and he was not disappointed. *I wonder if I can persuade her to give me a bed bath!*
"Right you are, nurse. Tell me, what's your name?" Mike never missed an opportunity to make contact with the opposite sex, in whatever state he found himself.
"Well, it's Joyce, but that doesn't matter now. You must take this medicine right away. Give it a chance to work on you."
"Never had a girlfriend called Joyce before," ventured Mike hopefully.
"You still haven't. Now give me your arm. I have to take your blood pressure."
"You can have both if you like. They'd fit round you a treat."

So he's one of the frisky ones, is he? thought Nurse Grey. *We get them now and again – even in a small hospital like this. I'll have to give him the Big Sister act.*

"Now you just be a good boy and behave yourself – and give me that arm before I call Matron!"

"Oh no, not that," Mike retorted in mock horror. "Anything but that. Just don't be surprised if my pressure's way up, that's all."

"We'll see." Nurse took his pressure and recorded it.

"You'll be glad to know it's pretty normal. The RAF ambulance is due here any minute, so be prepared to tear yourself away from us."

"Oh Nurse Joyce, do you always get rid of your patients like this?"

"Only when they're troublesome. But I'll see you before you go."

She then turned and made for the door with the words, "And by the way, good luck for your recovery." With that she was gone.

Mike settled down again, and looked around him. *Not a lot to see really. The ward looks clean enough, but the decor, such as it is, isn't very exciting. Still, what does that matter when you come here to be cured? Could be brightened up with a few pictures scattered about. Come to think of it, I've some rather attractive pin-ups in my room back at base which I might spare if they twisted my arm strongly enough. Thank Christ the old Spit didn't decide to catch fire. I'd have had it, and no error. She didn't seem to break up too badly either. That metal prop must have helped by bending and acting as a ski, just as George said it would. Mind you, puts a shock load on the engine front bearing, but you can't have everything. When Sandy hit the deck last week he had a wooden prop – blades broke off and saved his engine, but his Spit was a write-off, and he didn't look too clever himself either when we got him out. Life is just cussed.*

His thoughts were interrupted by the appearance of a RAF medical orderly. "Ullo, Sir. We've come to take you back to the holiday camp," he announced cheerfully.

"You know, you're all heart. But how am I supposed to get to the ambulance? Hop all the bloody way?"

"Not to worry, Sir, we've organised a wheelchair, but I'm afraid you'll have to put up with me driving it rather than a pretty nurse."

Mike adopted a resigned expression. "Into each life… but take it easy round those corners, my good man."

The orderly went back to bring in the chair, and the doctor returned.

"I want you to give this letter to your RAF doctor – it tells him what we've done for you here. We all wish you good luck."

"Many thanks, Doc, for your help – and your staff too. I'll come here for my holidays after the war. By the way, I can't find my mascot. Would be glad if someone could chase it for me. It's a small koala – answers to the name of Ozzy."

Under different circumstances, the journey back to base would have been quite pleasant, with a clear sky and wispy clouds above, and the land seeming to smile in the sunshine that had followed the rain, but nursing a sore leg and coping with not-too-soft suspension over roads bereft of routine maintenance during years of war gave Mike a somewhat white-knuckle ride. It was a relief therefore to turn into the camp gates, and be decanted into Sick Bay.

The RAF doctor read the note, and re-dressed the leg. "We'll try and keep you here if we can, but we're being pasted by Jerry regularly as you know, and I need to keep space here for any casualties. May have to transfer you to a Service hospital in the country somewhere."
"Thanks, Doc. I'd like to stay if I can. All my pals are here."
"We'll see. By the way, we tried to contact your parents, but there's been no reply so far. However, I was able to have a word with your brother, and he's coming over to see you."
"Thanks again, Doc."

Mike was just beginning to feel a bit more comfortable now. The dressing was giving support to his leg, and the much-needed sleep had given his body a chance to catch up and relieve stressed nerves. He settled back on his pillow, and was then addressed by the occupant of the next bed.
"Hello. How are you, friend? I'm Teo, from the Polish squadron here."
"Not too bad, Teo, thanks. My name's Mike. I've seen you in the Mess, haven't I?"
"That's right. You told me the names of some of the foods at dinner."
"Did I now? Hope I was polite!"
"It's not like we have at home, but good. Here, would you like to see this newspaper? I don't understand most of it, anyway."
"Many thanks, old man. If there's anything of interest, I'll point it out to you."

Mike went through all the pages, but it was mostly gloom and doom, with little to raise the spirits, and Teo seemed to be dozing off. Mike began to feel, optimistically, that his wound was not too serious, but realised that, if convalescence was prolonged, he would have a battle to fight boredom. He would not fancy doing jig-saws or reading novels for hours on end. And as for knitting balaclavas! *Stone the crows! Damn that for a lark. Must find something to do.*

His musings on these lines were suddenly curtailed when a bright face appeared round the door, and a familiar voice asked, "Are you sitting this one out, then?"

"Terry, you old sinner. Come and give us a kiss." Their hands met in a firm, meaningful, masculine shake.

"Terry, you did me a right good turn up there yesterday – or whenever it was."

For once, his voice was serious. He knew that his life had been saved by his companion's quick action, and that he himself would do the same for his brother officer should the situation arise.

"Glad to be of service, old son. Of course, you know it's going to cost you?"

"Not to worry, dear boy. I've already bought up shares in a brewery. We'll have a right thrash when I'm out of here. Incidentally, any gen on my kite, yet? Has anyone inspected it?"

"Yes. George sent out John to sort it. He came back about an hour ago."

"Cat E, I suppose?"

"'Fraid so. He said something about number ten frame being damaged. But many of the bits are serviceable, so they'll cannibalise it for spares."

"Pity really, but anything's better than finishing up shredded in a ditch somewhere. Did we lose anybody after I departed, stage left?"

Terry frowned. "Yes. Both Frank and Bill had it, sadly, and Brian is missing, but we saw him eject over the French coast so he might turn up as a POW."

"I sure hope so. He married not long ago, and his wife will be shattered. Poor old Frank – hasn't been with us long. And we'll miss Bill's comic songs in the bar, and no mistake."

"Yes. There's a lot of it about. Doesn't bear thinking about too much. War is an absolute bugger, isn't it, but we seem to be making it so far. Anyway, I'll have to push off now. Mustn't keep my public waiting, you know."

"Terry, thanks for coming. I'm already getting withdrawal pains without drinking partners."

"Right, old son, I'll just love you and leave you then. Got to press on with the performance. Besides, you ought to get some more shuteye while you can. Give us a shout when you're serviceable again."
"Will do. Bye."

* * * * *

The next morning saw the appearance of one of the medical orderlies, a rather chubby individual round-faced and genial.
"Hello, Sir, I'm Bert," he announced with a grin. "Come to take your order for breakfast. Will it be the smoked salmon as usual – and the kidneys to foller?"
"Not this morning, Bert," replied Mike, matching his humour. "I'll just toy with a cornflake."
Bert grinned. "Right-ho, Sir. I'll see what I can squeeze out of Cookie."
"Don't squeeze too hard – you don't know what you might get!"

Bert ankled off in the general direction of food, while Mike lay back and looked at the ceiling. *Well, what happens now? They tell me the leg's not too bad, so maybe it's a spot of convalescence after all, then back to the lists. No doubt the folks have been told by now. They'll probably come rushing over here to see what's left of me. Pity, really. There's no need.*
"Here you are, Sir," announced Bert. "I managed to wangle you an egg – yes, real live hen fruit – as you're aircrew."
"Well done, that man. Take an orange out of the basket."
"Do you mind if I watch you eat?" enquired Bert, with mock wistfulness. "Haven't seen an eggshell for ages. It's the dried kind for the likes of us."
"My dear feller – you can dip your bread in it if you wish – soldiers, of course."
"Only kidding, Sir. I'd rather tackle dried egg any day than do what you chaps do up there. Besides, they'd have my guts for garters if I was caught eating patients' grub."

Mike hardly had time to finish his meal when the chilling wail of the siren was heard, accompanied by the distant crump of gunfire. Instantly several airmen and WAAFs appeared, and one after the other the patients were led down the slope into the Sick Quarters' shelter. Then came the undulating drones of the German bombers, followed by whistles and bangs. The raid developed into a major attack. What was worrying was that these were becoming more frequent.

For Pete's sake, thought Mike. *Don't let me get written off now before I've had a chance for another crack at 'em.*

Chapter 2 – Matthew (1940)

MATT shut off the engine of his motorbike, and eased himself out of the saddle. It had been a long ride to the cottage hospital since early morning, particularly with no signposts to guide him, and his eyes stung with travel strain, heightened by concern for his twin brother's safety. They had always been pretty close, even though they were not identical, and their arguments and rows over the years had merely served to create mutual respect. They were dissimilar in many ways: Mike's light brown hair and eyes acting as counterpoint to Matt's darker appearance – the ready smile of the one contrasting with the more serious aspect of the other.

Mike was gregarious with a capital G, whereas Matt took longer to bond, albeit with a humorous response when he did so. Furthermore, Mike was clean-shaven – he had abandoned the traditional handlebar moustache (although it suited him) because it irritated him in his oxygen mask – whereas Matt supported a more modest version. They understood the Air Force to permit hair on the upper lip provided it was not cosmetically shaped, whereas the Navy stipulated either clean-shaven or a full set. Mike, however, did tend to leave the top button of his tunic undone – again, a fighter tradition.

Matt parked his bike and mounted the steps towards the chipped painted board indicating Main Entrance. The door creaked as he entered, and his nostrils instantly detected a faint blend of floor polish and some sort of disinfectant – quite different from the 100 octane and engine oil he was accustomed to. A middle-aged nursing sister came trotting towards him.

"Can I help you, Sir?" she asked, noting his officer's uniform under his leather jacket.

"Yes please, Sister. My name's Matthew Gregg, and I understand my brother Michael was brought in here a few hours back. Had a crash in his aircraft. We received a message saying he was admitted here soon after."

"Michael Gregg? Michael Gregg? Oh yes, that was that young flyer with the injured leg. He was brought in by some farm people who'd rescued him. Yes, we treated him and he had a sleep before the RAF came over to collect him. But wait a minute. Nurse Grey was on duty then – I'll get her to have a word with you. She can tell you more. Please take a seat over there while I get her. By the way, would you like a cup of tea?"

"Sister, I could murder one. Many thanks."

Matt sat down and tried to get his thoughts in order. *So that's it. They'll have got him in Sick Bay over at the airfield, so that will be my next port of call. Glad it's not far away. RAF bike saddles aren't the most comfortable things in the world.* However, he was in no way complaining. RAF officers do not normally have ready access to a service motorcycle, but his bike had been put at his disposal because of his frequent trips for aircraft crash inspecting and repair, and his boss had generously allowed him to use it for this emergency.

"Flying Officer Matthew Gregg?" A pleasant voice enquired over his left shoulder. Matt turned to see an attractive vision in nurse's uniform. "That's correct. Are you Nurse Grey?"

"Yes. I attended your brother while he was here. We didn't have him for very long before his colleagues came to collect him, but I'm glad to say that he was well enough to travel."

"How was he?"

"Not too bad, really. Mostly flesh wounds, it seems, but he'll be examined more closely by your people no doubt. He was very tired when we admitted him, but quite perky when they took him away. I'm glad we were able to help."

At this point, the tea arrived, which Matt accepted gratefully, and sank into the well-used sofa in the hallway.

"Tell me, are you the elder brother?"

"Not really. We're twins, although I was born a couple of hours before him, and I don't let him forget it!"

"That's the way to treat them," smiled Nurse. "I see you're both called Flying Officer M Gregg. Does that cause confusion?"

"Not a lot, although I can always point to Mike when anything goes wrong," laughed Matt. "Perhaps our parents were a bit thoughtless at the time. They could have called him Clarence, or something. I do hope he behaved himself while he was here."

"Can't complain," replied Nurse, tactfully. "By the way, he was concerned when he left that he couldn't find his mascot. Do you know what it's like?"

"Oh yes. It's a small koala character, name of Ozzy – given to him by one of our Australian cousins. He takes it everywhere. Do hope it turns up. Probably doesn't mean a damn thing, really, but you know how superstitious some aircrew can be – like the mariners of old, I believe."

"We'll certainly do our best to find it for him. Then perhaps we can send it along to his unit."

"Thank you very much. Well I'd better be off now and catch up with him."

"Of course. But have you eaten recently? Afraid we don't have a canteen, but I could probably rustle up something from the kitchen."

"Kind of you, but no thanks. I'd like to see what the medics have done to him back at his base. Er, no offence intended of course."

"None taken. You'll find a baker's in the village if you need a bite. They don't have much to offer usually but they'll probably find you something. Goodbye, and good luck, both to you and your brother."

She was nice, thought Matt as he started up and turned into the road. *I'd liked to have stayed a bit longer, but Mike has priority.* He cruised past a grassy field and noted a mobile crane stacking a damaged mainplane next to a rather forlorn-looking Spitfire fuselage into a Queen Mary low loader. *My word. Junior sure made a dog's breakfast of that one. Good job he got out of it as well as he seems to have done.*

On impulse, Matt steered into the adjacent entry yard of the farm when he noticed a boiler-suited figure wearing a woolly hat and a thoughtful expression, apparently on intimate terms with a tractor.
"Excuse me," he called, over the engine noise, "but are you the person who rescued the pilot of that aircraft over there?"
"Not me, mate, I wasn't here. It was the farmer. He's over behind that cowshed if you want to have a word."

Matt thanked him, switched off his engine and made his way to the cowshed, wheeling his bike. *No need to frighten the horses,* he decided. A slightly-built whipcord of a man looked up at his approach. He was wearing an ancient cap and a brown coat tied at the waist with string, and was holding a calf, which a land girl, encased in jodhpurs and a thick woollen jersey, was feeding with a bottle.

"I believe you're the ones to thank for your kindness to my brother for getting him out of that aircraft."
The farmer stood up, removed his cap and scratched his head. "Yes, but Doris here reached him first. Your brother, eh? Glad we were on the spot and able to help. Chaps like that deserve all the help we can give 'em."
"Is he all right?" enquired Doris.
"As far as we know. I'm sure he's sorry he rather spoilt your field."
"Well, I dare say that petrol will take time to clear, but it could be worse. Anyway, we're thinking of replanting there soon, so he just gave us a headstart. Give him our best wishes when you see him."
"I'll certainly do that. Thanks again. Bye."

Matt started up, and negotiated his way through the village. He noted the baker's location but pressed on. *Got to get to Mike as soon as possible.* Twenty miles of country roads led Matt to the gates of the RAF Station where he was checked by a duty guard. "Flying Officer Gregg? Yes, Sir, your brother was taken to Sick Quarters, but we've had a few raids lately so you might find him in the shelter underneath. Past the hangar on the left, then second turning right."

Matt found it, parked his bike, and made his way between the sandbags to the entrance. The medical corporal looked up from his desk and then, seeing the rank of the visitor, rose to his feet smartly. "Sir?"

"Flying Officer Gregg. I believe my brother Michael is here."

"Yes, Sir, he's just back. We've spent the last hour or so in the shelter, but have returned in the hope of a night's rest. Some hope!" he added resignedly. "In the ward there – second on the left."

And so he was – deep in conversation with a European-looking patient in the next bed.

"Hello, Junior," said Matt cheerfully. "How the devil are you? I've just seen your aircraft being disentangled from the local scenery. What are you doing damaging His Majesty's property – to wit, one vehicle, aerial, pilots for the use of?

Mike grinned widely. "Thanks a bunch for coming, old man. Must say I'm pretty chipper, one way and another. Just a slight modification to my undercarriage – but about a three-foot wide bandage with a couple of number nines thrown in should see me right, I shouldn't wonder. By the way, let me introduce you to Teo here. He flies with us when he's got nothing better to do."

"Pleased to see you, Teo. Don't let my brother bore the pants off you."

"Will not do that, brother. Have a nice talk. Please draw the curtain while I have a sleep." Teo had no intention of sleeping just then, but this was his attempt to give the brothers some sort of privacy in unavoidably crowded quarters.

"Now then," enquired Mike, "and how is my big brother?"

"Fair, but not improved by these antics of yours."

"What about Mum and Dad? Apparently the Adj couldn't get through to them."

"That's right. They're up at Brickhill at the moment. I managed to get Dad on the blower, and he's going to try to visit you here. As you know, Mum's coped with the bombing in London very well so far, but she's got a bit weary of it all so Dad got her up to the aunts so she could get some sleep and strengthen up."

High Firs, the Victorian country house in Brickhill, was reasonably large, particularly for the sole occupancy of two maiden aunts, although their constant aversions to each other meant that as much space as possible was favoured for them to pursue their lives separately on occasions following their frequent spats. Agatha and Belinda ('Aggy' and 'Baggy' in Mike parlance) were kindness itself to the rest of the family, and outsiders, but congenitally abrasive to each other, whatever the topic under discussion. The aunts had decided against living separately, even though there was a small cottage in the front garden because it brought in a useful rental.

"I'm so glad Mum's all right," said Mike, "although she must miss being in her own place, having to cope with those two dragons nattering away."
"Yes," said Matt. "She does worry. About you in particular."
"Aw, she mustn't do that. We're all in this giddy bloody circus together, aren't we, and the lucky ones will come through, no matter what."
"By the way, Mike, you know those farm people who pulled you out? I met them on the way here and thanked them for saving your miserable life."
"Oh good, I must track them down one day and stand them several beers."
"I also met the nurse at the cottage hospital. She said she'd look out for Ozzy, so don't be surprised if it turns up in the post."
"That's good of her. She's a stunner, isn't she? I was damn sorry to leave and come back to all these characters – nowhere near as good looking."
"Yes indeed," replied Matt thoughtfully, "she is good to see."

There was silence for a moment, broken only by a hearty voice approaching:
"Hello boys – glad to see you both."
"DAD!"
Smiles and handshakes all round. 'Dad' was a thickset man with a mature, quite handsome face topped with just a little hair. His hands were large, and his blue eyes twinkled.
"Mum sends her love. She's having a rest with the aunts at the moment – at least, she is when they stop arguing. A few days away from the bombing will do her a power of good."
"How did you get here, Dad?"
"Got a lift to Bletchley in the fishmonger's van. Then train to Euston, Tube to Victoria and a coach to the local town."
"Is the Tube still used as a shelter?"

"Very much so. People sleep all over the platform – there's only just room to get through them. Safest place, of course, except when it's subject to flooding."

"How's the ambulance business going, Dad?"
"Keeping us busy, but we've only lost one driver and a couple of vehicles so far. They've also made me chief warden of our street, so nights are quite exciting as well."
"And how's our big sister getting along?"
"Val's a real treasure. She looks after us a treat, and is learning cooking fast."
"Does she still work in that office in the City?"
"When she can get there after the raids. In any case, insurance is a reserved occupation. She also serves part-time in the fire service in the evenings. It seems everybody has to have two jobs these days."
"Well it all helps. Civvies and services seem to be hand in hand right now."
"That's very true," commented Dad. "Never known everyone mucking in as they do now. Before the war we hardly ever spoke to our neighbours, but now we share our Andersons, and know all the histories of each other's families. What's more, we tell each other when there's any meat arrived at the butcher's, or anything else in short supply. Sort of bush telegraph. Mind you," he added wearily, "I expect we'll all go back to our old ways when the war's over."

"How about the bombing in London, Dad. Is it very bad?"
"Well, it gets a bit naughty round our way sometimes. They're after the docks, of course, and Brockley is only a few miles away as the Kraut flies."
"What about the incendiaries? Do you have to go through the bucket-and-stirrup-pump routine?" asked Matt.
"Oh no, we don't bother with that. Just chuck some earth over it. That does the trick. It's the aerial mines that give the headaches. And they say that Jerry's putting whistles on his bombs so that they scream as they come down. Supposed to put the wind up us, apparently, but it's no problem. What's much worse is that some incendiaries have a delay fuse and they blow up just when you're putting them out! Damned unsporting."
"Those chaps who sort out the unexploded bombs," commented Mike thoughtfully, "they deserve the biggest of gongs. Must take enormous guts

to concentrate your thoughts when there's a load of explosives by your left ear."

"Are you eating OK – in London, I mean?" queried Matt.
"Mustn't grumble. A bit samey, but enough. Never thought I'd eat whalemeat, or whatever it is – and I'm sure that last bit of meat I had came from a horse."
"Was it that bit of saddle in your teeth that gave you the clue, Dad?"
"Idiot. Now is there anything you want, Mike? Anything apart from grapes or flowers of course."

And so the conversation continued, covering multiple aspects of a family joined by affection but displaced by war. Eventually, the doctor poked his head around the curtain. "Sorry to trouble you, gentlemen, but there's a warning of enemy aircraft crossing the coast. They might venture this way so we'll be moving our patients into shelter. I'd advise you to move away from here if you can. Just in case."

Following hasty farewells, Matt and his father had a brief word with the doctor who updated them on Mike's condition, and his probable transfer to a quieter environment.

"Let me give you a lift to the coach station, Dad," offered Matt. "I can't promise comfort, but I'll try and miss all the bigger bumps."
The station was empty when they arrived, but the clerk assured them that the London coach was on its way. "Goodbye, son. It looks as though Mike will be back in action before long. Pity really. We'd feel much happier if they gave him a desk job right now. But of course, he wouldn't want that. And you look after yourself, Matt. Expect you have some hairy moments too, don't you?"
"Nothing to worry your head about, Dad. In an OTU the only danger comes from trainee pilots mistaking the Officers' Mess for a Berlin factory and shooting us up. Cheers, Dad."

Matt swung away and steered his bike back towards the cottage hospital. He hoped to see that nurse again so that he could tell her that Mike was likely to be moved, and that it would be better if she sent the mascot to him, Matt, who could then send it on when Mike settled. (At least, that was Matt's excuse to himself. His subconscious had taken over from the rational, and instinct drove him to see that smile again.)

To his delight, Nurse Joyce was still on duty, so he was able to give her his address, and advice regarding the postage suggestion.

"Certainly I'll do that," she replied, "but did you have any luck at the baker's?"
"No, I didn't even try, I'm afraid. But they looked as though they're still open, so I'll see if I can get some joy from them now."
"I'm glad to hear that your brother is doing well. Good bye, then, and good luck."

Chapter 3 – OTU (1940 – 41)

Matt had not been back at his Operational Training Unit for more than a week when a small parcel arrived from the cottage hospital, containing the elusive Ozzy accompanied by the following note:

> Dear Flying Officer Gregg,
>
> We have just come across your brother's mascot, which is enclosed. It came to light when we checked through the laundry room – must have been tucked into his flying clothes. It was a bit soiled, but has responded well to hospital soap, and my hair dryer.
>
> We hope your brother receives it safely, and that his recovery continues.
>
> Incidentally, I am being transferred to one of the bigger hospitals, in Bedford, which is nice because my family live in a village nearby, so I'll be able to see them often.
>
> Kind regards
>
> Joyce Grey, SRN
> Staff Nurse

Matt lost no time in replying, first thing that evening.

> Dear Nurse Grey,
>
> It was so kind of you to take care of Ozzy as you did, and to send him to me. I am redirecting him to our aunts in Brickhill because Mike is likely to be moved about a bit, whereas we all

manage to meet up there together at times, so he can collect it then. He won't be needing it for any flying for a while.

Brickhill, of course, is not all that far from Bedford where we go shopping occasionally when the buses run. Please let us know your new address when you are settled, then perhaps we can all meet in town one day, and Mike can thank you personally for Ozzy's safe return.

Best wishes for your new appointment.

Matt Gregg, F/O
RAFVR

Matt found that engineering duties at an OTU were demanding in their own particular way – not quite as immediate and traumatic, he imagined, as those in a front-line squadron involving flak repairs, refuelling and re-arming, in addition to the myriad of servicing needs typical in anything to do with aviation. But they were demanding enough since their aircraft, being somewhat older and flown by pilots of, as yet, limited experience, required concentrated effort to ensure their complete airworthiness. Spares could be a bit of a problem at times because, of course, front-line squadrons had priority, so the OTU technical staff, at all levels, were repeatedly called upon to improvise.

Matt recognised the value of all this experience in shaping him for his eventual post-war career, and how lucky he was when compared to many others who were likely to return to a host of entirely different jobs in civvy street – their wartime experiences being of little use. This understanding helped to endure the sometimes sticky and unpleasant situations that invariably arose, particularly when loss of life was involved, or the weather was exceptionally foul. But, of course, all pulling together would mean there would be a civvy street to go back to. Which reminded him that he must make the effort to work more closely with Reg, his fellow engineer officer.

Reg was one of those backbone members of service units who joined as regulars before the war, intending to stay on for their complete careers.

Coming up through the ranks, with a thorough grounding in the practicalities of aircraft maintenance, gave them a hard-won confidence to tackle whatever technical problems might arise. They were sometimes unconvinced that an academically-trained engineer would have enough to offer an active service situation in which they were now embroiled. This had led to a somewhat uneasy relationship between them, plus a certain amount of rivalry, leading Reg to enquire how a theoretical background could be a lot of use in the very practical situation in which they found themselves.

"There are occasions when it can," Matt tried to reassure him.

"Name one," demanded Reg, quite reasonably.

"Well, this is fairly trivial, but we had to weigh one of our articulated long loaders before it was due to cross a rather dodgy bridge but, being long, it wouldn't fit on the weighing platform. So I instructed the driver to weigh the front part first – you know, the prime mover – then the rear part, and then add the two weights together to give the total. He just couldn't see that this would work. I tried to tell him that the weight on each set of wheels would depend on their distance from the centre of gravity, and the important point was that these two weights together must equal the total weight of the vehicle otherwise the bloody thing would either sink into the earth or take off and fly – but he just couldn't grasp it. My fault, I suppose, for not explaining it clearly enough.

Then there was the time when we were unsure whether the undercarriage of a Spitfire at the dispersal was locked down or not. I was asked for advice, so I said – 'Since the legs retract outwards, lash them together with some heavy-duty rope so they can't move.' Then everyone was happy. It's just a case of training to see the basic forces involved, and to act accordingly."

"I see," said Reg thoughtfully, "but what would you have done if it'd been a Hurricane with its legs retracting inwards?"

"I would have cut a tree trunk to size and jammed it between the wheels. It's just a matter of fresh eyes being useful."

Reg was silent for a moment. "All right, then. What else do your fresh eyes see that we haven't?"

"Well – and this is even more trivial – I've often thought about the batman polishing my tunic buttons. What a waste of effort."

"But it only takes a few minutes."

"True, but think of those minutes multiplied by the number of service people polishing away – and every day, too. Wouldn't that total time be better spent making munitions, or nursing the wounded, or even mending aeroplanes? And all that brass used for making buttons, and the button sticks for cleaning them. We're already taking people's saucepans and garden railings for the war effort, so why not the brass as well?"

"You've got a point there, Matt," Reg conceded. "Perhaps they ought to issue us with battle dress, like the Army."

"Huh," snorted Matt, "the Air Works' brass hats will never get round to *that!*"

But he was quite wrong.

Matt felt he might have made some headway in bonding with Reg, but that he somehow still had to prove to 'the regulars' that he had a positive contribution to make. Not that this worried him unduly. It was, after all, only fair, and it gave an incentive to 'go the extra mile' which would also help the war effort in its own small way, as well as his own future.

One morning brought an experience to Matt that he never forgot. A replacement Spitfire was to be flown in from the manufacturers by the Air Transport Auxiliary. Although a routine procedure, this time the aircraft was to be delivered to the OTU, for a modification to be installed, rather than direct to a squadron. Matt waited in the Watch Tower until clearance was given for landing, and then moved out to direct the pilot to the OTU's maintenance hangar. When the propeller stopped and the pilot jumped out to hand him the paperwork, Matt was completely surprised to see that the pilot was a slip of a girl! He had been so used to dealing with strapping young male aircrew that he was temporarily dumbstruck, but he pulled himself together and asked if the aircraft had performed satisfactorily during the flight.

"No real problems," came the reply. "The only thing is that she was slightly right-wing low, however much I trimmed it."

"OK, we'll check on the control lines."

"Is it right that you can bend up a few inches of the trailing edge of the starboard aileron?"

"Yes, that's right, because it forces the aileron down slightly, and so lifts the wing up. But we have to be careful and not overdo it, because once the edge is bent up it must not be bent down again."

"Does that reduce the speed at all?" She was obviously interested.

"Not seriously," replied Matt, "but would you believe it, the CO here once insisted that his aircraft be polished from nose to tail, with all the rivet heads smoothed down, and all the bumps knocked flat, and then reckoned that he got another twenty miles an hour out of it? Anyway, here's your Tiger Moth coming in to take you back. Thanks for the delivery. Happy landings."

Now why should I have been so surprised? Matt reflected afterwards. *Women are just as able and as gutsy to do a job like this as men.* Here was an example of a woman piloting a service aircraft (memories of Amy Johnson surfaced in his mind): they could also captain a ship, be successful doctors, researchers and follow a host of other activities. Although he reckoned that men were usually stronger physically, with wider shoulders and a firmer grip, it seemed to him that there were precious few jobs with which women could not cope just as well. Whether they could handle a heavy manual job like a steelworker's, or slug it out in the infantry on the battlefield, was not clear to him, but he'd heard that Russian women could probably do just that. This philosophy was a permanent feature of his life, and it always irritated him to encounter evidence of the subordination of womenfolk.

On the following evening, Reg was in the workshop supervising the completion of the modification, whereas Matt was updating the servicing records after doing his rounds, as Orderly Officer of the day, to check the dinner quality at the Airmen's Mess in the time-honoured 'Any Complaints?' fashion. The telephone rang.

"Is that the Orderly Officer?"

"Speaking."

"This is the Watch Tower. We've just heard there's a Whitley heading for us. It has engine trouble and will be making an emergency landing. Will you make sure that the runway is kept clear of any aircraft under repair, and stand by for any further instructions."

"Will do. We've already moved that Spitfire into the hangar. I'll stand by this phone until you clear me. OK?"

Matt threw open the office door and called out.

"Reg, there could be trouble on the way. A damaged Whitley's coming in. I have to stay by the phone, so could you do the honours please?"

"Right, I'll get the night crash squad into the 15-hundredweight and park near the runway ready to roll."

The normal routine of closing down for the day was now replaced by a buzzing urgency as the fire wagon and ambulance took up their emergency stations, and Reg and his crew drew up behind them in case they were needed. Before long, the distinctive nose-down shape of an Armstrong Whitworth Whitley emerged through the twilight with one engine out of action and the other struggling. Everybody tensed as the pilot departed from the usual go-around procedure and aimed directly for the runway.

In she came, and it looked at first as if the pilot's skill was going to be amply rewarded. But the damage was not confined to the engines because, as soon as the weight of the machine began to transfer to earth, one undercarriage leg buckled and the aircraft slewed over across the airfield, through the neighbouring farm, and slammed into an isolated farmhouse where it promptly burst into a ball of fire. Aircrew could be seen jumping out and racing for cover, whereas the rescue vehicles were desperately forcing their way through the crops.

Matt waited anxiously for news. *Did they all get out? And what about the farm people? Is there anything Reg can do?*

It was late at night before a weary Reg returned and reported to him. "The crew all survived. Few injuries. But those poor farm people. Bloody awful. The woman and her baby were in bed. Didn't have a chance. You could see the iron bedstead through the flames. Her husband turned up while we were there. Poor sod. He'd been to the local pub, and when he was told what had happened he just sat down on a stone and wept. It's bad enough to see any people written off, but I hate it when kids are involved. We had to leave then because apparently there are bombs still on board."
"Reg, you need a drink, and I am buying you one, so no arguments."

The next morning found the whole station in a sombre mood. Breakfast was taken without the usual banter and, in fact, conversation as a whole was in very short supply. Matt had been up at first light to view the scene from the hangar roof and note the extent of the damage. He was shocked at the sight. The swathe cut in the crops revealed that the stricken aircraft had first made a straight path well clear of the house, but had then snagged on some bushes and been redirected unerringly to its tragic terminus, where a group of men and vehicles could now be seen.

Of all the lousy, rotten, bloody awful luck. Why couldn't it have gone straight on? Why did innocent people have to die? Where's the sense of it all? Matt had to walk about a bit before he went in to breakfast. He found himself sitting next to Reg. Neither of them spoke much, but both felt that they had something of an understanding in their shared anger and grief.

Matt was alerted by a thwack on his shoulder by a rolled-up newspaper. It was S/Ldr Chase, the Chief Engineer Officer of the unit.
"Matt, I want to see you. My office. Ten minutes."
"What have you been up to?" asked Reg, with the ghost of a smile.
"God knows, but I hope there aren't any more shocks today. I've just about had a bellyful. See you."

Squadron Leader Robert Chase was a seasoned regular officer who'd just about seen it all. As a Halton 'brat' he had been steeped in aircraft engineering from early rotary-engined flying machines to modern front-line fighters, taking in light bombers, twin engines, wooden fuselages, and a number of experimental aircraft that didn't quite make the grade.

"That was a bloody awful event last night, Matt, wasn't it, but there's precious little any of us can do about it right now. But I want to speak to you on another matter. The CO has been on to me about this new business of engine handling to economise on fuel. Not in battle, of course, but when the pilots are cruising or patrolling or watching over shipping and that. There's a course on at Derby, but he says he can't spare any of his flying staff to go, and he certainly can't send all the trainees on it. So I agreed to send you. You've always shown an aptitude for engines, haven't you? I recall your suggestions when we had to replace the mountings in that

Mustang. So your job will be to mug up all you can on the course, then come back here and teach the trainees whenever the weather is too bad for flying. How does that grab you?"

"That sounds great, Sir." Matt was only too glad of the chance to concentrate on something other than the tragedy of the previous night, and he had to admit to himself that, despite the overall approach to aeronautical engineering in his university degree, it was the engine that fascinated him more than the airframe structure or the aerodynamics.

"Right. Get along to the Adj to fix up details. And report to me when you get back. I expect you'll be away about a week."

Matt saluted, and made his way along the corridor to the Adjutant's office. All the arrangements had been made. He was to join a class at *Rolls Royce*, and be accommodated by them. He would have use of his service motorcycle, and was to hand over his duties to his fellow officer, F/O Reginald Hall, before he left.

"I understand," said Matt, looking ahead, "that I'll be giving lectures to the trainees when I come back. Is there some sort of lecture hall we could use?"
"Not a chance. I'm afraid everything's in use for one thing or another. But we could square up one of the flight crew rooms a bit. That might do the trick. Case of having to, really."

Well, first things first, thought Matt. *Must get all my gear together, fill up the bike, and prepare for off. Mustn't forget plenty of pencils and paper either. It'll be like going back to school, but it'll be just great to land at the birthplace of the Merlin. Must keep my ears well and truly pinned back. They'll have a mass of information there.* After a pause, his thoughts continued: *I wonder if I could then swing it enough to get myself to Bristol, and learn more about the radials and sleeve valves and stuff. Perhaps, but one thing at a time. I'll just see if I can leave a message for Mike about this. He'll be highly chuffed by the news. Derby, here I come.*

Chapter 4 – Watershed (1941)

The journey to Derby was mildly exciting for Matt, not through any adventurous encounters on the way, but from the heady prospect of extending his knowledge about a very high-performance piece of aero machinery. He was beginning to understand more deeply than before how much he appreciated the joy of learning, and of understanding the principles and not just the practices alone of the engineering world.

As the miles steadily passed, Matt reflected on the pattern of his life that had led to this moment. He recalled that sister Val, being a girl, was the first of the three of them to mature mentally, and consequently had won her way to a high-class Secondary School. He himself failed to matriculate, but won an award of sorts and was sent off to a Central School. This did not please him because he had visions of becoming a draughtsman, drawing aeroplanes all day long, whereas this school specialised in commercial studies, and drilled him in the intricacies of shorthand, typewriting and bookkeeping. However, he must have done something right because he was appointed Head Boy. Mike had shown no academic prowess whatever, but breezed through an Elementary School where he excelled in all the sporting activities they could offer (and probably some they didn't, but Matt had no wish to pry!)

Because he had shown such a keen interest in the sciences, Matt had gained entry to a University Diploma course in Aeronautical Engineering, but failed the Intermediate Exams and had to resit. This, he recognised later, was the best thing that could have happened to him, because he buckled down and managed to devise a system of studying that suited him perfectly for preparing for, and passing, written examinations. As a result, he was transferred from the Diploma to the BSc Degree course and, in Part 1 Finals, did reasonably well. In Part 2, he reached Honours level. Some day, he had concluded, he would see if he could publish his study system, and so help others.

(Details of this study scheme can be found in Appendices 1 and 2)

One of the attractions of aviation was the fact that it appeared to him to be the most advanced branch of engineering being practised at that time. (This was before the onset of Nuclear Engineering.) He would never forget the introduction to that very first lecture: Aeronautical Engineering is like ordinary engineering, but more difficult; and later on when they were being taught Propulsion: The aero engine is the athlete of the species.

Even though his interest in aircraft as such was being superseded somewhat by that in aero power plants, Matt congratulated himself on choosing a technical discipline of such high quality.

He smiled as he recalled his first weeks as a young engineer officer in the OTU. Standard RAF practice with such rookies was to detail them for crash inspecting, the theory apparently being that they were unlikely to do any more damage to an aircraft already out of action. So they had to learn to categorise damaged aircraft as either repairable (Cat AC) or fit only to be reduced to produce (Cat E), and thus gain invaluable experience in discovering which components were likely to fail on impact or from enemy action, or even from just plain metal fatigue. Even so, Matt was delighted that his greater love, for engines, was being requited in this way.

On reaching Derby, he turned into the engine works, and was directed to the Training Block. He was shown a motorcycle parking spot, then to his room in the accommodation building. "Dinner will be in about half an hour, over the road," he was told. "You'll meet all the other members of the course there. See you then."

After a welcome wash and brush-up, Matt strolled over to the dining quarters to be greeted by a mixed group of men from a variety of air forces all busily chatting away over jugs of ale. The conversation was heavily technical, with – as the saying has it – 'the hangar doors wide open'. Matt allowed himself to be absorbed by the crowd as he gravitated towards the bar and ordered a beer.

"Matt!" called a voice from the edge of the throng. "How are you doing?" It was the recognisable tones of Ted Saunders whom he'd met on the introductory course at Henlow. Matt turned sharply in order to meet up with his old mucker, narrowly missing the drinking arm of a tall young Australian Air Force officer in the process.

"Sorry, old son," apologised Matt. "Hope I didn't spill any."

"She's right," was the smiling reply. "Plenty more where this came from."

Matt and Ted then updated each other with events since being pitchforked into the harsh world of service life from the cradle of their training days, and exchanged gossip about their fellow students from the Henlow course. At dinner, Matt found himself sitting next to the Australian. "So sorry I nearly pranged you back there. Let me buy you a drink after dinner."

"No worries. Tell you the truth, I'm not all that chuffed with English beer. It's warm! We always have it chilled back home – even in the remotest pub in the bush. There'd be a riot otherwise. I'm Dennis Walker, by the way. From New South, training to be a Flight Engineer."

"Nice to meet you, Dennis. I'm Matt Gregg. Have you been over here long?"

"Just a few months. We've been moved about a bit, so I haven't had much chance to make friendships. It's good to come on a jolly like this – just to meet people for one thing."

"Know what you mean. I expect we'll both make some good contacts while we're here this week."

The following morning saw everyone assembled to await the beginning of the show. The course was well organised, having been run on several occasions, with a set of printed notes laid out on each desk. The lecturers appeared, introduced themselves and then, having briefed everybody about the domestic arrangements, the first lecturer went straight into it. "This course is all about Engine Handling for Fuel Economy. It's based principally on the Rolls-Royce Merlin, but applies generally to most supercharged main aero engines. The text my colleagues and I will be preaching throughout is this: High Boost. Low Revs. But we won't just leave it at that. We shall go through the various steps to show how we arrived at it. The notes in front of you are provided in case there are any points you miss in the lectures. Also, we shall give you a chance to tour the manufacturing and assembly bays so that you'll see how these engines come together. We'll finish up with a question-and-answer wash-up session so that we can all sort out any bugs that may have arisen."

Then followed hour after hour of instruction into carburettors, superchargers – two speed and two stage – intercoolers, altitude power, boost control, constant-speed propellers, internal combustion, mixture strength, fuel consumption, friction, engine construction and a host of related matters. Matt wrote furiously. He didn't always have time to check whether everything was in the printed notes, so he did his best to record all the most important bits to make certain. This was when his earlier instruction in shorthand proved invaluable, but he had the sense to know that he would have to transcribe it within hours otherwise much detail would be forgotten.

Matt knew intuitively that he could not present all this new information in an acceptable fashion until he had had an opportunity to digest it and then reproduce it in some edited, cohesive way that he himself could understand fully. Only this would suit the needs of budding pilots who were not necessarily technically equipped prior to their flying training. Although this rewriting kept him busy each evening, he deliberately found time to discuss the course with fellow class members. Matt always maintained that as much knowledge was gained that way as in the lectures themselves. Matt and Dennis found it particularly fruitful to compare notes and bounce questions off each other. Dennis had a refreshing, clear-minded approach to life, and although blunt at times, was without malice. They began to bond so well that, on impulse, Matt found himself saying, "Look, if you're ever round London at all – south east that is – don't hesitate to call in at our home. Dad used to have some relatives in Townsville, and I know he'd like to meet you and hear your news about Down Under. Here's the address." Matt thought it was unlike him to dish out his family address to new acquaintances, but he was happy to make an exception for such good types as Dennis.

The learning effort was intense, and they all welcomed the break to visit the factory. This proved equally exciting as the lectures, but in a different way. Matt's initial observation was of the notice hanging from the factory roof. Everyone, service personnel and civilian alike, was used to placards exhorting maximum war effort, and the need for constant security – 'Be like Dad. Keep Mum,' was one that intrigued Matt. But this one was so punchy, so concise, and so apt for a high-tech production line that Matt could not suppress a smile of respect. It simply read: 'Good enough is not good enough!'

There were acres of workbenches and assembly rigs tightly packed into islands of different activities – machine tools churning out components at one end of the factory, with assembly taking place progressively towards the final build of complete engines at the other end before they were loaded into large packing cases and put aside for the crane. To the outsider, the whole operation looked chaotic, but its success was underlined by the shortness of delays between crates being filled and then removed for transport. Matt knew that his brother would love to see this – It'd give him even more confidence than he already had in his engine. Matt was somewhat surprised, and greatly impressed, by the large number of women workers involved throughout. *What a damn good job they're doing,* he thought. *I bet the suffragettes of old would be glad to see this – about time women's skills were recognised.*

Inevitably, the appearance of a group of young, presentable servicemen was enough to generate a few wolf whistles from some of the younger women (and some of the older ones too!). Matt smiled. *Nice to be considered worthy of being whistled at,* he thought. *And some of them are really attractive too with their headscarves on, and their faces lit up with smiles.* It reminded him of Nurse Grey back at the hospital. *I wonder how she's getting on in Bedford,* he mused, but was then immediately occupied by a demonstration of how the entry blades were located onto the supercharger disc. Next came the static test beds. Here an engine fitted with a calibrated propeller was roaring away with the unmistakable throaty growl of a Merlin, almost painful despite the issue of ear defenders to each onlooker.

The final wash-up session proved one of the most valuable aspects of the course, as they invariably do. Dennis wanted more explanation of engine performance in hot climates, and the elimination of sand from engine intakes on take-off. Matt, on the other hand, pursued a line of enquiry regarding the meaning of the 100/130 grade designation applied to the petrol used in aircraft practice.

"The figures relate to the anti-knock quality of the petrol. That is, 100 octane number in so-called 'weak-mixture' cruise, and 130 performance number in high-power rich-mixture conditions of take-off, climb and combat." *Intriguing,* thought Matt, *but it's not quite enough. I'll try to follow this up before leaving.*

The course broke up after lunch and, having wished Dennis farewell, Matt made a bee-line for the lecturer concerned. "Thanks for answering my query on the 100/130 grade, but do you mind if I delve a bit deeper?"

"Fire away, er, it's Matt, isn't it?"

"That's right. Look, I think I understand the first figure – we were taught at university about anti-knock rating of petrol in a single-cylinder engine with variable compression. If I recall, you keep raising the compression ratio by winding the cylinder-head down until detonation occurs. Then you keep repeating the test with blends of the reference fuels heptane and octane until you find one that detonates at the same compression ratio as the fuel you're testing. You can then quote the octane number of your test fuel as the percentage of octane in the matching blend. Have I got it right?"

"More or less, but I would add that it is *Normal* heptane, and a particular version of *Iso* octane. Normal means the straightforward unchanged molecule, whereas an isomer is a rearranged molecular structure. And the engine is known as the CFR. Also, I personally don't like the term 'detonation' since I think this is something much more energetic that takes place in a laboratory combustion tube, for example. To my mind, the better term is 'Spark Knock'."

"Right, I'll remember that. So that explains the label '91/96' that appears on the fuel tanks we use for our light training aircraft. It's 91 octane number when fuel-weak, and 96 octane number fuel-rich. Yes? Also, the figure 100 means that our standard operational fuel is as good as iso-octane itself at fuel weak, right?"

"Correct."

"So what's with the 130 figure? You can't have 130 per cent, can you?"

"True, but what has happened is that today's aviation petrols have been so developed that they are superior to iso-octane at the rich condition. So what they've done is to add variable amounts of tetra-ethyl lead to the iso-octane reference fuel to improve it even further, and to compare the resulting increased power you can get with it before it knocks. This enabled them to devise a range of so-called 'Performance Numbers' reaching about 160 with one-and-a-half ccs of tetra-ethyl lead added to each litre of iso-octane. So '130' means that you can get 30 per cent more power out of the engine at rich mixture, without knock, than you could running on pure iso-octane."

"I see," said Matt, genuinely intrigued. "What a fascinating business, isn't it?"

"I think so. If you like, I can fill you in on the history of the TEL search. It's unique. Would you like to hear it?"

"Would I not? Please go ahead."

"Well, it all started round about the 1920s. A lot of the early work was done by Ricardo in this country, and people like Kettering, Midgley and Boyd in the States, with variable-compression engines and experimental fuels made up by blending the best components available at that time. One of the things they found was that some materials added to the fuel had very profound effects on combustion, even in very small concentrations, and they called these 'dopes'. But I like to think that the real breakthrough occurred when they noted that kerosine knocks much heavier than petrol. They thought this might be due to the lower volatility of kerosine, and when they looked out of the lab window they noticed that one of the bushes outside was able to live through the winter. Ah, they thought, this could be due to the fact that it was coloured red, and so could absorb sunlight more easily. They promptly decided to dye their kerosine red, too, and see what happened to the knock. Having no red dye in the lab at that moment, and with the shops shut, they opted for some iodine and, to their delight, found that the coloured kerosine did not knock so heavily.

"However, when the shops opened again and they repeated the experiment with proper red dye, they were dismayed to find no improvement. After some thought, they realised that it was the iodine itself that had been acting as a chemical anti-knock dope. They then got down to this research in real earnest. Their aim was to test as many of the chemical elements as possible to check their effectiveness. They took a piece of wood and drilled a number of holes in it, each hole representing one element of the periodic table. Into each hole they placed a peg to show by its length the anti-knock effectiveness. In this way, they went through a number of materials like selenium and tellurium, both of them giving such bad odours to the skin and clothes of the testers that I believe there was talk of strike action! However, the sizes of the pegs guided them unerringly towards lead, and they found its tetra-ethyl compound to be the best of all."

"Why did the lead have to be in the form of tetra-ethyl?"
"Because you can't slice up bits of metallic lead and throw them into a tank of petrol. They just wouldn't dissolve. What you have to do is surround each atom of lead with hydrocarbon radicals so that this new molecule is compatible with the hydrocarbon molecules of the petrol itself. Lead is tetravalent, so it joins up with four ethyl radicals giving tetra-ethyl lead, TEL, appearing like of a ball of hydrocarbons with the lead tucked inside it. Cunning, eh?"
"Sure is, but why is TEL so effective?"
"Well, knock is now understood to be due to the spontaneous ignition of the unburnt mixture as the flame progresses across the combustion chamber, and the theory is that the lead combines with oxygen during the main combustion process to form lead oxide, PbO_2. Then when the unburnt mixture is about to ignite spontaneously and create knock, it breaks this compound down into PbO and loses so much energy by doing so that it cannot ignite. But the PbO rapidly re-oxidises to PbO_2, and repeats its Good Samaritan act over and over again."

"Oh I see, but did they have any snags in those early days?"
"I'll say. They recognised that lead is poisonous, and adopted very careful methods of handling it. Also, they found that, although the lead did a marvellous job in the engine cylinders, it poisoned them too because it couldn't get out via the exhaust valves afterwards. But they soon found a compound that would combine with the lead after it had done its job, and allow it to evacuate with the exhaust gases. The best scavenger they found was ethylene dibromide, but then, of course, there was a shortage of bromine, and they were forced to extract it from sea water. But that's about as much as I know. If you want to find out more, you'd better ask a chemist."

Matt felt elated. This was just the sort of information that inspired him, and stimulated him for much more study. In fact, it hit him like a thunderbolt that *this* was going to be his real specialist area for the rest of his career. Starting from an overview of an aircraft as a complete entity, as he had so far been involved, his interest had steadily concentrated on the means of propulsion, i.e. the engine within the aircraft, as indicated by his presence on this course. However, it had now focused, permanently, he was certain, on the fuel within the engine. He would, he resolved on the spot, transform himself eventually into a fuel technologist, but it would be a long haul, and he realised that he must go back to dealing with the whole aircraft again for the remainder of the war, because that was his current responsibility. In any case, it would all be excellent background experience for him, and he also wanted to learn much more about the leadership aspects of handling staff and engineering projects.

Matt prepared to leave with feelings of great excitement and determination. This event was to prove nothing less than a watershed in his life. The trials and tribulations of the future could never be so awesome with this Grand Plan in his mind. His destiny had been revealed that day with crystal clarity.

Matt took one last look at the factory buildings before he left. There seemed to be acres of industrial roofing, surely a tempting target for enemy bombers. He marvelled at the way aircraft production seemed to be coping despite it all. The Ack Ack and the barrage balloons must be doing a great job plus, of course, the night fighters. But how on earth was this all possible? It would be several years before he learnt of the part played by the motor industry in the Shadow Scheme of wartime aircraft production.

Chapter 5 – Brickhill (1941)

Matt found the drive back through the afternoon countryside to be invigorating, matching his mood of exhilaration and confidence. His musings ranged over many of the things he had gleaned from the course. One in particular was the fact that, by replacing the long exhaust pipes that were fitted to pre-war aircraft like the Hawker Furys and Harts with stubby rearward facing pipes for each pair of cylinders, not only was the exhaust back-pressure reduced but some additional thrust was given to the aircraft. This just hadn't occurred to him so clearly before. But of course, anything that added to the propeller airstream by ejecting fluid backwards would react to give thrust forwards. Only small, of course, but every little helps. Many people looked on a propeller as corkscrewing its way through the air, which is why it used to be called an 'airscrew' until one misinterpreted order from Stores resulted in the appearance of four fully-kitted up members of an 'aircrew'. He also realised that things like coolant and oil radiators, by adding energy to their cooling airstreams, would also give thrust. Matt had just started to give much more serious thought to the concept of Jet Propulsion.

When he arrived back at his unit, he found the following letter from his mother:

Dear Matthew,

I hope you are well and getting enough to eat. I am pleased to say that the rest has done me good, and I should soon be returning to London to look after Dad and Valerie.

The woods here look very nice, and the rhododendrons were beautiful this year. We go for walks when we can, with Rex enjoying every minute, although he is not quick enough now to catch rabbits any more. We visit the village shops nearly every day and are managing to live quite well.

"Well actually.... I requested an air**screw**!"

As you know, Aunt Agatha's leg is very bad. It doesn't seem to improve so the doctor referred her to the hospital in Bedford. We went there yesterday, and she's been advised to wear support stockings. We'd told her time and again to do this, but it looks as if she really must do it now. While we were there we met a very nice young nurse by the name of Grey. When she saw our names she asked if by chance we had a Michael and a Matthew in the family. It turns out that she was the first one to look after Michael after he had his accident. Isn't it a small world? She struck me as a very nice girl, and I've no doubt that Michael noticed that too – you know how he is? I'm so glad that he has recovered, and hope he doesn't meet any more trouble. When he rang the other day to tell us of his return to his unit, he mentioned that you are going to some lectures somewhere. Do hope you enjoyed them.

I am still worried about your Gran – she is still living in Battersea and is really too old to have to put up with all the London bombing. I'd like her to come up here for a rest, or to stay here for good. But Grandad, of course, won't budge – you know how stubborn he is.

Your Dad says how pleased he is that America has joined us now, but it is terrible that so many lives have been lost at Pearl Harbour. We are concerned for the people living in the Old Dairy opposite. They have lost their only son on the Hood. Dad says he is glad that the Bismarck has been caught at last.

We look forward to seeing you when you can get some leave, and all send our love

Mum.

Matt busied himself in settling into the service routine, reporting his return to the Adj and S/Ldr Chase, and recounting some of his recent experiences to fellow officers in the Mess. When evening came, he sat down to start correlating all the notes, leaflets and impressions he had gained at Derby, and to meld them into a logical continuing narrative charting the progress of fuel-air mixture from its entry to the engine to its eventual ejection from the exhaust system. He decided to tackle it from two angles: the first being a comprehensive story including as much theory as he could remember from university days, and the second being the extraction of those items which alone were required for his forthcoming presentations to the trainee pilots. He recognised the urgency of the situation because the weather might clamp at any time, at which point he would be commanded to take the lecture rostrum and regurgitate the engine handling mantra.

Despite his excitement about the task in hand, he somehow couldn't get started. It was all whirling round in his brain, but he felt like he was searching for the free end when trying to unwind a ball of string. He walked round the room for a while, and stepped outside to take in some cool evening air. Then it dawned on him. That mention of Joyce Grey in his mother's letter had set a subconscious chord vibrating, masking other thoughts and emotions. For some time now he had vaguely realised that he needed the comfort, the beauty and relief coming only from female companionship, but had been too preoccupied with work to be fully aware of it, and so had put it to one side. But now he recognised that Joyce had appeared repeatedly in his thoughts, and even in his dreams. He resolved there was only one thing to do, and that was to make contact with her, and to get himself over to Bedford somehow as soon as he could manage it. He could then ascertain if there was any chance of a real friendship – or not. It had to be one way or the other, and if he got in first, who knows, he might win her affections before Mike got a look-in.

Having thus cleared his thoughts, at least for the time being, he returned to his writing table, picked up his pen, and weighed into his lecture notes. When he felt he had done enough for one day, he allowed himself the luxury of writing to Joyce – as follows:

Dear Joyce,

I was so pleased to hear from my mother that you had both met, and you are now settled in to the Bedford scene. As you know, we come over there when we can – to shop, of course, not always to bring Aunt Agatha's leg!

I expect to have some leave before long (if the RAF can manage without me for a while!), and I wondered if it would be possible for us to meet so that I can take you somewhere for a cup of tea to thank you on Mike's behalf. I am glad to say that he has now recovered well enough to soon return to duty. I trust you are well, and enjoying your new job.

Look forward to hearing from you.

Sincerely

Matthew Gregg

A few days later saw an event of a dramatic and tragic nature. Aero engines are meticulously designed and manufactured artefacts but, like everything else, subject to faults in their old age. A trainee pilot was circling the field after take-off when one block of the Vee layout of his engine apparently warped sufficiently to permit ethylene glycol coolant to leak into the exhaust stream. The resulting trail of white smoke led the pilot to conclude that he was on fire. The shock was sufficiently strong to persuade him to leap out and attempt a parachute drop at that low altitude. His parachute had little time to open, and he was killed. Had he but followed instructions, he would probably have been able to effect a safe emergency landing. Meanwhile, Matt and Reg were left to organise the removal of the crashed aircraft from the nearby field.

Matt was grateful to note the easier relationship he now had with Reg. The fact that Matt had been selected for the Derby course, and had recounted the major points so generously to Reg, was a matter for mutual respect.

The following day saw a reply from Joyce:

> Dear Matthew,
>
> Thank you for your letter. Yes, I would be pleased to see you. They tend to shift us about to different departments here, but at the moment I am working in Accident and Emergency, so when you come, look for the A & E sign. The full address and phone number are shown on the hospital leaflet enclosed. If I'm not around, my colleague, Brenda, will take the call and winkle me out.
>
> I hope your aunt continues to improve, and that you are all well.
>
> Best wishes
>
> Joyce

This pleased Matt no end. He now had the means of direct contact with Joyce, and he intended to make good use of it. Soon after he received another letter from his mother:

Dear Matthew,

I'm writing again to tell you that the house next to Gran and Grandad has been blitzed. Fortunately they are not hurt. An empty house nearby has been offered them but I took the opportunity to encourage Gran to come up here to Brickhill. I'm so pleased to say that she has agreed. What's more, Grandad has agreed to come as well – very reluctantly, of course. But he says that Jerry built these old houses, and now that Jerry has knocked them down, he feels free to leave. We shall miss all those vegetables and fruit he used to grow in that tiny garden, but he'll probably enjoy the garden here.

We're now busy making arrangements to move them both, with all the bits and pieces that they will need to bring. Thought you'd like to know.

Your loving

Mum

PS Michael has just phoned. He thinks he can manage to get his old car going so that he can help the move to Brickhill. Kind of him, isn't it? Love. M.

Matt resolved to try to see if he could help with the move. The weather seemed to be holding up so he might not be needed to lecture for a while. Perhaps he could wangle a 48-hour break, or even a 72, if he spoke nicely to the boss.

S/Ldr Chase seemed to be in a reasonably expansive mood that day – the CO's congratulations on the rapid turn-round of several aircraft may have had something to do with it – and he approved of a break for Matt. "After all," he said, "I expect you found that course pretty gruelling." Matt made no response. Silence can be golden rather than rabbiting on about how enthralling he had found it. "Cut along to the Adj then – and ask Johnson to get me another coffee."

After leaving the Adjutant's office, Matt had an inspiration and called in to see the sergeant in Motor Transport.
"Morning Sarn't. I don't suppose you have anything going Northampton way, have you?"
"Fraid not Sir but, – wait a minute, there is a 3-tonner leaving for London in about 15 minutes. We could give you a lift there if that's any use."
"Brilliant. Thanks a lot. I'll go get my gear."

Matt rushed around like the proverbial blue-tailed fly, and reappeared at MT ten minutes later with his small kit lashed to his gas mask. The miles trundled by at a stately pace, and the corporal driver found it easy to talk to Matt on the way – he was not like some of those toffee-nosed officers he'd encountered.

Sensing Matt's interest in fuels, the corporal ventured a question. "I know you de-fuel aircraft if they fail to take off, but what do you do with the petrol? Would it suit an old banger like this?"
"Well it might," replied Matt, "but the fact that it's better than ordinary motor fuel would be no good to you because you can only get the benefit of the higher octane fuel if you stressed the engine more to where it would otherwise knock. In other words, if you could supercharge the engine, or increase its compression ratio, or even advance the spark a little, you could get more power but without knock. Just putting a better fuel in and no more does no good at all. You have to make things more severe to get the extra performance. Also you may find that the additives in aviation petrol would deposit on the plugs."

"Thank you, Sir. I've often wondered. So there's really not a lot of point, apart from using the stuff up rather than wasting it?"
"Exactly. By the way, do you go anywhere near Euston on your route?"
"Funny you should say that, Sir. I shall be going right by it in about half an hour," he said with a grin.

There was quite a wait at Euston for the Bletchley train, because of bomb damage to the lines just outside, but Matt eventually found himself in a corner seat, watching the passing back gardens, sticky-taped windows and Anderson shelters. The so-called connecting train at Bletchley did not live up to its name, so Matt decided that he would walk the rest of the way. However, his luck was in, and a passing motorist took pity on him, as frequently happened with service people. "Have you heard?" asked his helpful companion. "The Ark Royal has been sunk."
"Oh God, no," exclaimed Matt. "That's another kick in the teeth for us all. When is it all going to end?"

Suppressing his disappointment, Matt arrived at *High Firs* to find Mike's old banger (affectionately named Mabel) gracing the driveway, and several of his family chatting excitedly in the kitchen. Rex came over to him as fast as his elderly pins permitted. As a retriever puppy, he had revelled in the rough play doled out by Mike, who was his blue-eyed boy as a result. He never tired of tracking down a thrown stick and savaging it to death at Mike's feet. However, in later years he had become more sedate, and transferred his preference to the gentler touch of Matt when he was around.

It was then that Matt noticed a significant change in Mike's uniform. "Hey, you've got a second ring on your sleeve. When were you promoted to Flight Lieutenant? And why didn't you let us know?"
"Aw well, it only means a little more beer money coming in each week – not enough to write home about."
"Well, congratulations anyway. You're coming out with me for a celebration drink later."

The next few hours were occupied with furniture rearrangement, wiring up of bedside lamps, emptying and restocking of linen cupboards and all the usual activities of hastily revised accommodation arrangements. By nightfall, all concerned were eager to crawl into their beds and enjoy a noise-free night of peaceful sleep.

After the finishing touches during the next day, Mike indicated that he was ready to proceed to some nearby tavern to partake of a celebratory jug or three. "There's that lovely little pub in Milton Keynes village. You know, it's got a gamekeeper's bar and a poacher's bar." Mike had always had trouble deciding which was the better, and his researches in that direction were ongoing. "The only thing is, I'm just about out of petrol, which is why I'm leaving the old banger here. Still," he added, as the result of a sudden brain wave, "we could always borrow the aunts' bikes. They wouldn't refuse a couple of splendid chaps like us." And they didn't.

And so it came to pass that two RAF officers could have been observed traversing the Buckinghamshire countryside mounted on "superior ladies' bicycling machines (as supplied to the intelligentsia – easy terms arranged – J Morgan, Propr.)" complete with curved crossbars, front-mounted baskets, and cord netting to prevent skirts interfering with the rear wheels, a horrendous thought. An edifying spectacle, indeed. "If Hitler could see us right now," opined Mike, "he'd surrender on the spot."

Once ensconced in the window seat behind two large jugs of the local brew, the brothers began to feel that, for once, they could relax and enjoy these precious moments together. "Glad you were able to bring the old folks up from London, and all their gear too. How on earth did you get the petrol, Mike?"
"Don't ask, dear boy, don't ask. Suffice to say that I have a friend who has a friend who, for the right price, will provide the essential gravy. Mind you, he charges like the Light Brigade, and I could only get enough for the one-way journey, but in any case I'd like the old girl to be stored in safety in the garage here. I'd feel easier in my mind that some Kraut is not going to let off his machine guns on her back at camp. I'll get a train in the morning."
"You know, Mike, we ought to get her up on blocks to save her tyres – and also take the battery out and keep it on trickle charge. I can do that for you before I go back, but why not drive over to catch the train from Bletchley and let me bring her back here – you've got enough petrol for that, haven't you?"
"Most kind, dear Sir. Thanks a lot."

Matt then launched into a summary of the events at Derby, and of the recommendations for engine handling. "You know," said Mike thoughtfully, "some chap from Group came round preaching the same gospel, but I'm not sure I got the full strength of it until now. Well, I'm not likely to run out of

fuel now before getting back to base. Mind you, if I do, I'll come back to you and complain!"
"I'll remember that. Same again?"
"Yes please, and get a packet of crisps or something if you can."

Matt moved over to the bar for refills, then returned. "Sorry Mike. Ale yes, food no – they're fresh out."
"No food? Things have come to a pretty pass – that's what things have come to."
"Has your leg been playing up at all, Mike?"
"Not really, but I'm still intrigued about those coloured lights in my eyes."
"Oh, that will be the petrol fumes, Mike. They do that sort of thing, according to our doc. By the way, have you got your new aircraft yet?"
"Yes, it's another Spit. I love those kites – that curved wing planform is so sexy. Mitchell got it just right. Mind you, the Hurricane is another great bus. I've flown them in the past, and they do a famous job. That wide undercart is great for taxiing. Mind you, I had a taxiing accident once in one of those. Mangled the wing tip of the CO's plane.
"Was he mad?"
"Mad? Well, let's say there was a marked shortage of the milk of human kindness for a while, but I managed to help him see off a couple of Heinkels later, so he forgave me."

"Are they keeping you busy now you're back?"
"Busy? Cripes, I often don't know whether I'm Arthur or Martha – and I don't even have time to look! How about you? Are you still bustling about earning an honest crust?"
"I'll say. While you pilots keep bending our aeroplanes we don't have a chance to pack our hammers away. Now tell me, Mike, have you flown any twin-engined jobs?"
"Oh yes. The old Anson, of course, for back and forth bus journeys to the satellite airfield, but I also rode in a Beaufort. And have you seen the Beaufighter? – A really vicious looking job. Glad I'm not the enemy when that's around. By the way, I know they laughingly call you an engineer, Matt, but do you ginger beers ever get a chance to fly? Have you ever flown a kite?"
"Well, yes and no."

"What the hell does that mean?"

"Well, a pilot friend took me up in a trainer, and handed over the controls for a while. I seemed to be coping fine when I suddenly realised that I'd worked on these aircraft as a student during a works period from university."

"So?"

"The fact is that I wasn't at all happy with the work I did. It was soldering tubes and brackets together to support the fuel tanks, and all I managed was a very gobby job indeed."

"But surely all these things are inspected before they're thrown together and wheeled out into the great blue yonder, aren't they? Well, I hope they are anyway." Mike added, thoughtfully.

"Of course they are, but I only thought of that later. Stupid of me to panic. But quite frankly, I find myself much happier when handling a boat. I think that's what I'll get up to when this lot is over. Incidentally, I've decided where my future is going to be – I'm aiming for the fuel business, probably research and teaching in some college or other. I've always been interested, as you know, but it all came clear to me at that Derby course. Have you decided what you're going to do, Mike?"

"Not yet, but I suppose I might apply for a job in an airline. I could describe myself as 'Driver, Airframe' in the good old RAF way. By the way, I was in the Stores the other day and was highly chuffed to see that jerries, the guzzunder type, were listed as 'Pots, Chambers Officers, and Pots, Chambers, Airmen'! One had wings printed on it, and the other a crest, I think. Would you credit it? I wonder what the WAAFs have?"

"I think we'd better draw a veil over that one."

"Talking of the future, I did have a great idea the other day, and chewed it over with Teo. We reckon that, as soon as the war is over, all these Spitfires we've been aviating in will be unwanted, and available all round the country for next to nothing. If we can buy them up cheap, then store them and keep them serviced, we ought to be able to use them for air displays and films and things. If Teo joins me, I thought we'd call ourselves 'Spit and Polish'! – Do you get it?"

"I do. But why stick to Spits only? Surely there are many other types of aircraft that might have a future. Mind you, you'd have to find quite a bit of cash up front. Where is that coming from – and don't look at me, I shall need all I can scrape up to get myself organised in civvy street."

"That, dear boy, is the rub, but what did the good Mr Micawber always do? Why, he waited for something to turn up, and boy, am I good at waiting?"

"How is Teo, by the way?"
"Quite perky, really, although he reckons he's lost most of his family back in Warsaw. One thing that delighted him is that his only sister managed to sneak out somehow and get to Britain. He showed me her photo. Boy, is she a cracker? He wants me to meet her, and he's not the only one, I can tell you. Now she's here, he doesn't want to go back home. Not surprising, really. Now it's my turn to buy you a pint. Hey, bartender. Would you be so kind as to replenish these jugs please, and tread them well in while you're at it, there's a good fellow."

The return journey was negotiated without incident, both men becoming convinced that the bikes ran more smoothly with the help of copious ale despite the lack of landing lights in the growing darkness.

The next morning saw Mike taking his passenger to the station. "You know, there's not a lot to write home about in this place, Bletchley, is there? One or two good shops but no central feature like a clock tower or town square. Mind you, there seem to be a lot of people using the trains, and quite a lot are women, too."
"Probably there's some industry nearby," commented Matt, with thoughts of Derby in mind.
"Did you know that Val had an interview here somewhere? They were interested in her studies of German at school. Don't know what it was all about. Anyway, Matt old boy, thanks for looking after Mabel. Incidentally, I think we ought to take Dad out for a drink one evening – perhaps next time we meet here. Don't you agree?"
"Great idea. We'll do that. I'm sure we can resurrect his old bike at the back of the garage. I'll have a go at it as soon as I get the chance."
"Right. I'll be in touch when I get back. Hope the old folks will be happy here. Cheers, and look after yourself."

Matt watched the train vanish steadily into the distance, and gave an involuntary shiver, hoping they would see Mike again soon. He returned to Brickhill, and jacked the car onto blocks, as promised. He also removed the battery and put it on charge. There was Dad's old bike – dusty but unbowed. As he walked from the garage, the neighbour called over the fence.

"How are you Matt, and how's that brother of yours?"

"Nicely, thank you. And how are you coping with retirement?"

"Never been busier. I miss the company of my mates at the paper works, but it's good to be up here in the quiet rather than on the Thames riverside. Seemed an ideal target there for the Jerries, and you can't hide a river on a moonlit night. There's one interesting thing that happened recently," he warmed to his theme. "We get our water supply from artesian wells on the site, and one morning after a night raid we started to find petrol in the water. We complained to the oil company's tank farm next door, and it appeared that a bomb went right through the floor of one of their tanks but didn't explode, so all the petrol leaked out and contaminated our water supply. So the oil company fitted up some sort of filter system, extracted the petrol from our water, and put it back into another tank. Now that's what I call smart."

Matt thanked him for his story, which he would add to his own collection. As he had already learnt at university, it's always useful to have something unusual or amusing to add to a lecture – helps the medicine go down.

Having played his part in helping to move the old folks to a quieter environment, Matt now felt free to follow his own needs. He phoned Bedford Hospital, and Brenda replied to the effect that Joyce was heavily involved at that moment with a traffic accident casualty, but that she would be free at five thirty that day. Matt took a hasty lunch, and then made quickly for the bus stop. The bus seemed to relish its presence in the Bedfordshire countryside, and be in no particular hurry to reach its terminus, but eventually it did, and Matt disembarked without further delay. He had a quick recce round the town to see if there was anything like a reasonable tea shop that was open and had something to offer its customers, eventually homing in on the restaurant of a department store. Having completed his research, he aimed for the hospital's A & E department just before five thirty.

A tall, fair-haired nurse with a straight back and deliberate movements approached him. "Can I help you?" she enquired.
"Yes please, I'm looking for a Nurse Grey. Joyce Grey. My name is Matthew Gregg."
"Oh yes. She's in the side ward. Hang on, and I'll get her. I'm Brenda, by the way. We spoke on the phone. Just a moment."

She disappeared into the side ward and, reminiscent of a weather house, out popped Joyce. She had had a demanding day, and looked somewhat flushed which, thought Matt, made her look even prettier.

"Hello Matthew, I'm so glad to see you. It was good of you to come all this way."
"My pleasure," replied Matt, trying to keep his voice under firm control. "Incidentally, I'm known as 'Matt'. I hope you don't mind giving up your time to a lonesome serviceman."
"I'm looking forward to it, Matt. Just give me about ten minutes and I'll change and join you in the hallway. Have you met Brenda yet?"
"Yes. She kindly tracked you down for me."
"Of course. Nice, isn't she? We've become very firm friends. She wants me to be her bridesmaid next month. Right. See you in a bit."

In retrospect, on the bus home, Matt concluded that this had been the best evening he had spent as long as he could remember. Their conversation ranged over everything and anything, and the pleasure they derived in each other's company became so obvious that the waitress judged them to be a honeymoon couple. Matt was captivated by Joyce's every word and movement, with a growing realisation that this was going to be the woman for him. Joyce, for her part, warmed to the steadiness and thoughtfulness of her good-looking escort, enjoying his humour which was a touch more seriously based than the outrageous comedy of people like Mike, likeable though they were.

Reluctantly, Matt had found that it was time to see Joyce back to her quarters if he was to catch the last bus for Brickhill. They had stood at Joyce's doorstep for a few precious moments. Their eyes met, and silently conveyed the messages they wanted to tell each other. Eventually, Matt had cleared his

throat and said, "You know what this evening has meant to me, Joyce. I want to have many, many more like that. I want you to look on me as a very special friend – and to come out with me as often as we can manage it. Will you do that, Joyce Dear?
"You know I will, Matt. And I'll write to you often, as well. Do look after yourself, my dear."

She had kissed him lightly, and he had folded her in his arms and kissed her very firmly on the lips, and pressed his cheek to hers. He had then raced for the bus station, at a height of about three feet above the pavement according to his reckoning, and just made the bus as it was about to tackle the return journey. On this occasion, the journey duration did not trouble him as he sat in the front seat glowing. The bus could venture around *all* the local villages it wanted to as far as he was concerned. He had met Joyce again, he had confirmed to himself that he wanted her absolutely, he had shown her his feelings very clearly, and she had responded even better than he had dared hope.

Matt Gregg was one very happy man.

Chapter 6 – Kismet (1942)

Matt's feelings post-Bedford were a mixture of elation and determination, with a background warmth of glowing confidence. This seemed to support his appetite for work. The flow of newly qualified pilots from the initial training schools showed no signs of abating, so OTUs were kept busy bringing them up to operational standard. Routine servicing and maintenance of the aircraft was interspersed with the odd mechanical failure thrown in for good measure.

Then one morning a letter arrived from his father:

> Dear Matt,
>
> I'm glad to hear that you and Mike were able to help out so nobly with the removal to Brickhill. It was so much better for my parents to travel direct like that than having to keep changing trains, especially with all that luggage.
>
> I thought you'd like to know that we had a visit from your Australian friend, Dennis, the other day. He seemed so pleased to 'get his feet under a family table' as he put it, and we enjoyed his company a lot. In fact, he took quite a shine to Valerie, and she to him it seemed. She insisted on seeing him to the station, and on the right train for Charing Cross. I was able to tell him how grateful we are for that food parcel we've just received from that lady in Melbourne. It's good to know that Empire people are thinking of us at this time. I said we would be writing to her shortly to thank her.
>
> It's good to hear that Monty is making some progress now in Africa. We could do with some good news after losing Ark

Royal last year, but poor old Malta is getting a bashing. Hope Mike doesn't get sent there.

Things have been a bit quieter in London lately, I'm glad to say, but we get the impression that the airfields are getting attacked pretty regularly. Hope you stay safe.

Cheers for now,

Dad

A letter also arrived from Mike confirming that he now had an even later mark of Spitfire than he hoped for, but poor old Teo had crash landed and suffered fairly extensive burns, so he was being patched up at East Grinstead. Apparently, his sister had arrived to help care for him and he, Mike, hoped to see both of them before long. He advised Matt to ensure that he kept his bowels open at all times! Ever thoughtful!

Matt had a rather hairy experience one morning when starting his tour round the airfield to check on his aircraft under repair at the various dispersals. A pilot had just attempted to take off in a Tiger Moth to bring back a guest to a dining-in night. Being in a light aircraft, he aimed to use a cross-field short take-off, but misjudged it, and the kite stalled, coming to rest against the cab of a petrol bowser and then bursting into flames. He was killed instantly, and the flames could be seen licking round the blackened bones of his forearms. The one or two airmen nearby were rooted to the spot in shock but Matt, not having actually witnessed the accident, remained in a calmer mood and realised instantly the danger of a burning aircraft jammed against a bowser full of aviation petrol, with the hand brake forced back into the locked position because of the impact. The octane quality of the fuel meant nothing under these circumstances, but the fact that it was volatile petrol meant that it was highly flammable, so Matt raced round the back of the bowser, tied a rope to it, and tried to drag it away. It wouldn't budge, but he was driving a Jeep at the time, and remembered that these vehicles have a four-wheel drive capability, which he duly selected, and gradually the bowser responded and moved out of harm's way. By this time, the fire truck had arrived, so Matt continued with his airfield tour. It was many years later, after conducting tests on the fire risks involved with fuel tanks, that Matt realised that, had the fire reached the bowser fuel, it would not have just caught fire – it would have exploded, and probably shot the bowser into the air! Ignorance can be bliss, sometimes.

Matt, together with two of his aircraftmen, then experienced another dodgy situation. The three of them had been sent out to follow up a report of a crash involving one of their aircraft. Even though they could have driven themselves, for some reason a WAAF driver had been detailed to take them by truck to the crash site, which was located in hilly countryside in North Wales. Progress was uneventful until they approached a hairpin bend leading down a hillside flanked by a steep drop into the valley below. Quite unexpectedly, the driver uttered a scream and clapped her hands over her eyes. Instinctively, Matt grabbed at the wheel and tried to keep the truck on the road, but he over-steered, and the truck flipped over on its roof and careered for a further hundred yards or so in this manner.

Matt was aware of a mix of airmen and toolboxes in the back of the truck, and of petrol seeping into his scarf. Fortunately, there were no injuries or fire, and passing motorists rapidly extricated them and offered assistance. Leaving his crew to recover, Matt accepted an invitation from a motorist to proceed to the site of the aircraft crash, where he was fortunate to find one of the Group officers already in attendance. Noting that the pilot had been lost, and the aircraft completely written off, the officer eventually returned Matt to his crew, and then to base.

Matt found it distasteful, but necessary, to attend the subsequent hearing when the WAAF driver had to face a charge. He hated to see a female in trouble but, if they were attempting a man's job, they had to face man's discipline. However, he salved his conscience by pointing out, quite honestly, to the female officer in charge that he had no complaints whatever about the driver's competence prior to the accident, which is why it had come as such a surprise.

The day duly arrived when the British weather clamped, as the British weather invariably does, and the call came for Matt to occupy the trainees' time with the engine handling story he had gleaned from Derby. The fact that he had prepared his notes thoroughly, and practised delivering them time and again in his room, even in front of a mirror, gave him the confidence to feel that he was suited to the task ahead, but the thought that this was his first public presentation of, to him, knowledge gained only recently added a frisson of nervous energy. This was a good thing, he reckoned, as it prevented any danger of becoming complacent or over-casual in his approach, especially in front of qualified pilots who had the job of actually handling the engines in flight.

On thinking back to his university days, he realised that he had subconsciously learnt something about presentation by observing the lecturers in class, with their differing techniques. Some had merely breezed into the lecture theatre, fixed their eyes on the clock at the back of the room, droned on interminably with little change in tone, then breezed out again with no indication of sources of further information. The better ones, on the other hand, greeted the class on arrival, explained what they were going to cover, then did so with varying intonation and stressing, maintaining eyeball contact with each student in turn, then summarised what they had covered, and directed them to follow up references before departing. He also recognised the value of some humorous asides, because the facts associated with the humour had never left his memory. This was the technique he was going to use, he decided as he picked up his notes and strode purposefully to the makeshift lecture room.

The trainees eyed him somewhat quizzically. They already had flight experience behind a main aero engine, and were not entirely convinced that a non-flying member of the ground staff had much to teach them.

This was just the challenge he needed, and Matt opened his notes on the table. He had vowed that he would refer to them as little as possible, relying on his memory and addressing his audience with his head up, rather than mumbling away with his eyes down on the table. He also decided to always use the term 'we' and not 'you'. This would reinforce the fact that he, like them, was also a student of engineering. Without further ado, he went in for the kill: "Good morning, Gentlemen. Today we are going to refresh our memories of the various ways in which we can coax the required amount of power out of an aero engine. We will then seek the best way to do this, and explore the reasons behind it. First, let's remind ourselves of the layout of a typical aero engine."

Matt unfolded a large simplified diagram he had prepared of a piston engine equipped with a carburettor and supercharger at the rear, and a propeller shaft at the front. "We have two master controls over our power output. The first is a throttle lever connected to the butterfly valve in the carburettor intake that controls the boost pressure through an automatic unit. The second control is by a lever connected to the constant-speed unit that works by automatically adjusting the pitch of the propeller blades, as shown in this second diagram. So there is an infinite number of combinations of boost pressure and engine speed for any given power output. When we need high power for take-off or climb, we aim for the maximum severity that the engine will stand – and the instruction books lay this out clearly. In fact, for combat, as we all know, we can push the throttle through that piece of wire to go into 'boost over-ride', but it has to be reported later, and that wire acts as a tell-tale. This is because the engine can stand only a limited period of this very high stress. But our job today is to select the best combination of boost and engine speed for cruising economy, so that we can maximise range and endurance when the need arises."

Matt then went on to demonstrate that, since the inlet pressure is *restricted* by the carburettor butterfly and then *boosted* by the supercharger, these two components tended to fight each other, and it therefore made greatest sense to keep the butterfly as far open as permissible. On the other hand, since engine friction *increases* with rotational speed, it was advisable to keep this down. Taken together, this gave the optimal combination as High Boost and Low Revs, and Matt showed various graphs to illustrate the point quantitatively.

Matt sensed that the message was getting across, and that any audience misgivings were evaporating. For his part, although Matt had already noticed at university that an audience learns much about the characteristics of the lecturer – did he change the tone of his voice, avoid eye contact, cough frequently, walk about the podium, turn his back on the audience when explaining images on the screen, etc – he now began to realise that the lecturer also learns a lot about his audience by recognising their level of attention from the absence of fidgeting, scuffling and whispered asides to neighbours. He already knew, also, the old joke about a lecturer who did not mind when a member of his audience stared in dismay at his watch, in disbelief that so little real time had passed when it had seemed like ages, but he did object when the member shook the watch vigorously, listened to it ticking, and looked at it again! Matt was gratified to note that no watches had been consulted on this occasion.

He was quite well pleased with his performance, although he recognised a couple of weak points that would need attention, but he was also aware, from his academic background, that there was a lot more underlying theory that should be sorted out before he had mastered the depth of understanding that he sought. Hopefully, he'd have a chance to study this some time when things were quieter. And on this occasion he was right.

Even so, Matt well knew from talking with Mike and the trainees at the OTU that pilots have an awful lot to think about, particularly in battle, so perhaps it might be possible to link the throttle and engine speed lever together to give a correct combination of boost and revs at all conditions of flight – or was this too complicated? Events proved this not to be so, and the Single Lever Control was, in fact, introduced soon after.

Flushed with modest success, Matt found his way to the bar that evening, to be met with congratulations from a number of his colleagues who had been able, and had made the effort, to attend his presentation. Matt was glad to note that these included Reg, who shook his hand warmly. The evening became quite convivial, and Matt felt that he was now beginning to get both his private and professional lives organised in the way that he had hoped. Then a steward came up and addressed him.

"Excuse me, Sir, but Squadron Leader Chase would like to have a word with you. He's in the billiard room."
Matt felt even more pleased. Was the boss going to congratulate him, too? This was really great.
"Sit down, Matthew." Something in the boss's face and manner drove a sudden chill up Matt's spine. "First, congratulations on your performance today. A good job well done. But now I'm afraid I have some bad news for

you. Very bad news. The Adjutant received a message today that your brother, Michael, has had an accident. I decided that I wanted to give you this news myself. There's no way of softening this, Matt, so I'm very sorry to say that he has been killed. In action apparently. It was witnessed by his fellows so there is no hope that he has survived. You have my condolences – from all of us, in fact. Your parents have been informed, of course, and you'll no doubt wish to join them as soon as possible. I suggest that you get to bed as soon as you can, and see me in the morning when we'll fix you up with some compassionate leave."

The boss shook hands very firmly, and showed Matt out of the room where Reg was waiting, having been alerted beforehand. "Matt, I am so very, bloody, sorry about this. I've been instructed to see that you are OK, and get back to your quarters to rest. Would you like a drink – a strong one – before you go?"

Matt shook his head. He was completely incapable of speech. The shock seemed to have gripped his stomach, and left him feeling light-headed. No Mike. No more Mike, he thought blindly. All those years came rushing back to his memory. Ever since the day of his birth there had been Mike to play with, argue with, fight with occasionally, and eventually grow up with as affectionate, mutually respecting, twin brothers who had even entered the same branch of the armed forces. Mike with his ready wit, wicked humour and downright cheerfulness. Matt's thoughts immediately turned to his parents. What would they say? He must get in touch right away, and go and see them. He must head for the phone.

His fingers trembled as he dialled *High Firs*. His father answered with a thick voice clouded with emotion. "Dad, I've just heard. It's terrible news for all of us, isn't it? Dad, how has Mum taken it?"
"Remarkably well, really. I think she was fully prepared for something like this to happen eventually, and there's a sort of calmness about her now that it has happened. It's as though the worrying she's been going through has slipped away. Gran is very distressed, of course, and so is Granddad – and the aunts. Even Rex seems to sense that something is wrong. Any chance of seeing you soon?"

"Yes indeed. The boss is giving me some leave, and I aim to arrive some time tomorrow."

"I'm pleased about that. Quite frankly, we need you here with us just now – very much so. We'll see you tomorrow then. Goodbye, son, and thanks very much for calling."

Matt then tried to contact Joyce, but she'd gone to the operating theatre with a patient so he had to leave a message with one of her colleagues.

In the morning, Matt dragged himself out of bed after a troubled night, and forced himself to go through the usual routines. Breakfast did not appeal, but he sensed the need to eat. S/Ldr Chase was firmly kind. He explained that he had visited Group HQ recently, and heard that instructions were to be issued shortly to the effect that technical officers were to become adept at handling a wide range of transport vehicles, from motorcycles to 3-tonners. Although not stated, the reason was obviously part of the planning for the forthcoming invasion of Europe. Consequently some Jeeps had been earmarked for the Engineering Section, and Matt was to take one over and become fully competent with it. Furthermore, the Equipment Officer had indicated that there was an aircraft spare part awaiting transfer to Thurleigh in Bedfordshire. Was Matt up to handling all this?

Matt assured him that he was, and then prepared for his journey. On the way, he found that he had to positively stop thinking about Mike, otherwise he ran the risk of losing concentration and straying across the road. He had already accustomed himself with the Jeep, despite its left-hand drive, and was now well aware of the four-wheel drive mode. His first concern was that it had no ignition key, just an on-off switch, so security might be a problem. Also, he would not have been very happy in a combat situation with the petrol tank just under the driving seat! Later, when it came on to rain, he found manual windscreen wiping every few seconds to be tedious in the extreme and he vowed that if ever he was issued with a Jeep of his own, he'd make sure he found an electric wiper from somewhere, and fit it.

Matt delivered the component to Stores at Thurleigh, then proceeded towards Brickhill. He was tempted to stop in Bedford en route, but realised that his priority was to reach the family as quickly as possible. Also there was no secure place to leave the Jeep. Arriving at the family home, he shared their grief and did his best to help out in a quiet, practical way. There was so little one could do or say – merely try to carry on and give what comfort he could.

Dad, who had managed to travel up for a day or two, came in to tell Matt that he was wanted on the phone. "Hello Matt, my dear. It's Joyce. I got your message about the terrible news. Matt, I'm so dreadfully sorry. Mike was so full of life - it just doesn't seem possible that we shall never see him again. I had only just returned last week from Brenda's wedding to Norman, and was beginning to feel so pleased for all of us! I do hope that you and your family are coping well at this time. Please let me know if there's anything I can do – anything at all."

"Thank you, love. Yes there is. Please let me see you this evening. It's urgent, believe me."

"Of course, darling. I'll be off at five thirty again, so I'll see you as before. Goodbye for now, and please, *please*, look after yourself."

Seconds later, the phone rang again.

"This is Terry Watson. I flew with Mike a lot, and we were close friends. I'm so very sorry this has happened. I don't know what the official message to you was, or how much they've told you."

"Only that he'd been killed in action."

"Well, I was there and saw what happened. Do you want to hear it?"

"Yes please. Everything."

"Well, we were mixing it with some Heinkels over the Channel, and Mike was hit, with smoke starting to trail out. He was clearly in deep trouble, but didn't bale out. Perhaps he couldn't. The next moment he dived onto one of the Heinkels and literally sliced its tail off! The aircrew could be seen parachuting down, but Mike's plane simply exploded and disappeared into the sea. I'm sorry to say this, Matt, but there won't be any remains to bury. There's nothing left."

"Yes, I understand."

"There's one other thing, though. Mike didn't have his mascot with him at the time! We had to scramble pretty smartly that morning, and he must have overlooked it. It was lying on his dressing table when I looked into his room. Hope you don't mind, but I thought you might like to have it, so I've taken the liberty of holding onto it for now in case it got lost when his effects were sent to you. Did I do right?"

"You certainly did, and I'm sure we're all grateful to you. Please send it along to here. And thanks again for calling me." As he replaced the phone, Matt wondered if Mike realised that he didn't have Ozzy with him.

Matt caught the earliest bus he could for Bedford, leaving the Jeep parked safely behind *High Firs*. He could not assemble his thoughts into any coherent order on his way there. His urgent need was to have the warmth of Joyce by his side, and in his arms. He walked rapidly to the hospital, and tried to concentrate on the nearby shops before he felt he could reasonably wait for her inside the A & E without disrupting her work unduly.

Joyce was well on time and, after kissing him, took him up to her room where she changed and then joined him to cuddle on the sofa. They tried to find words of comfort for each other, but Matt found himself seething with a mix of utter misery, nervous exhaustion and raw desire. The only possible comfort in life for him at that moment was the feel of her soft rounded body under his hands. He sought to undo the tiny buttons of her blouse, driven by an instinctive urge for her bosom, repeating actions that must have taken place repeatedly over centuries of human existence, but his trembling hands could not manage the task. What happened then gave him one of the greatest thrills of his life because Joyce opened her blouse herself, slipped the bra off her shoulder and held his head against her naked breast. Matt's lips then searched for, and found, her nipple, and he drew in deeply as if back in babyhood. He then laid his head against her, and burst into deep sobbing, releasing all the tensions and emotions that had been building up within him over many months.

Eventually, he reached a stage when he realised clearly that he must either consummate the relationship, or stop right there and return to something like normality. He chose the latter – but not through lack of courage to take the ultimate step. On the contrary, it was because he had sufficient strength of control, and because he had argued many times to himself that what he really wanted in his private life was sex with love, not just sex alone, and only when he had demonstrated real respect for his partner by waiting for marriage beforehand.

He went to the bathroom to freshen up, while Joyce readjusted her clothes and waited for him. She was equally pleased that they had had the self-control to wait to fulfil their love for each other.

Matt returned, and they sat together again, holding hands.
"Joyce, my love," said Matt, "that was the most difficult decision I've ever had to make in my life. You know now just how much I want you, and that I respect you enough not to ask too much right now. But I can't hold back much longer. So I'm asking you to marry me as soon as we can arrange it – and I'll go on my knees right now if you wish."
"No need to do that, my love," smiled Joyce. "Of course I'll marry you, Matt, and I'll be very proud to be your wife. So now it's a matter of making arrangements, as you say. When Brenda was married the other day, she jokingly said that I could use her wedding dress if I wanted to. I might just take her up on her offer – wedding dresses are not too easy to find at the moment."
"Right. What time are you on in the morning?"
"Not until noon – I'm on a late turn, why?"
"I shall be over as early as possible, then we'll find an engagement ring. This may be a bit of a shock for the family after Mike's death, but life must go on, and it may ease their sadness by giving them hope for the future. They'll have something lovely to think about for a change, bless 'em. I must now go and catch that wretched bus. Goodnight, my sweetheart, and thank you so much for loving me."

With that, he was off. Where there was tragedy, there was also hope.

Chapter 7 – Bonded (1943)

Matt stared gloomily out of the coach window as he left London for his unit. Once again the OTU had been generous in giving him a brief respite, this time in order to attend the memorial service for Mike. He was so pleased that Joyce had managed to overcome the travel restrictions and attend; it was a good job that she had to visit a London hospital at the time to collect some samples. He only wished she could have stayed longer, but even a few hours were precious, despite the sadness of the occasion. How kind it was of Dennis to escort her from and to the station – good job he had managed a couple of days off.

On thinking back over the service itself, Matt was satisfied that the vicar had repeated all that he, Matt, had advised him regarding Mike's background. In fairness, there was no other way since there was a large hole in the vicar's knowledge of Mike, due to the latter's contributions to apple scrumping, conker tournaments and various other sporting activities in preference to attendance at Sunday School. Nevertheless, Matt felt a touch frustrated. After all, he knew as much as anyone about Mike's somewhat colourful past, and he could have stood up and recounted this with a depth of feeling based on personal knowledge. Matt was already recognising the urge and ability to express himself firmly on matters that concerned him deeply, and the recent experience of lecturing to the trainees was strengthening his confidence to do so.

As he walked through the gates back at his unit, Matt seemed to sense an atmosphere of urgency. Nothing tangible, but people moving about with more than the usual speed, and an apparent heightened sense of purpose. He was not surprised, therefore, on reporting to the Engineering Section, that Reg approached him to announce that the boss wanted to see him immediately. "It rather looks," said Reg wryly, "that the balloon is about to go up soon." Somewhat puzzled, Matt reported to the office of S/Ldr Chase.

"Ah, there you are, Matt. I trust all went smoothly in London."

"Thank you, Sir. Yes, quite smoothly. I understand you wish to see me."

"Correct. You know, of course, that words like 'Invasion' and 'Second Front' are being banded about a lot now. We shall be going in to Europe all right – no question. Only way to end this war. Well, I've been up at Group HQ a lot recently, and there are major changes in the pipe-line. Some top brass will be visiting units soon to spell it out. I gather that many of us will be offered choices for various jobs, rather than just posted willy nilly. Very considerate of them, I'm sure!

"The overall plan is for squadrons to be arranged into groups of three, four or five and called 'Wings' – and they will conduct their own daily servicing with a small group of fitters and riggers. Then each Wing will be backed up by an RSU, that is a Repair & Salvage Unit, that will handle the more extensive work. Underpinning all this will be a Forward Repair Unit of some sort to cope with the really big repair jobs. The whole lot will be grouped under the Second Tactical Air Force (2nd TAF). I don't know whether any of these prospects grab you, but if so let me know as soon as possible so I can put in a word for you."

"Well, thank you very much, Sir. The RSU sounds great to me. I like the thought of the mobility, and the level of work involved. How about this OTU here, though?"

"Oh, that will probably fold. I don't suppose they'll need to keep training so many pilots if this tea party is as successful as we hope. Right, then. Keep alert for those interviews from Group, and carry on with plenty of driving experience. The Adj is fixing up with MT to give us access to all sorts of vehicles, and we'll have to practise travelling in convoy, too. It's a good deal slower than driving a single vehicle, you'll find. They measure speed in just a few miles every two hours. That's about it for now. Any questions?"

"There is one thing, Sir. I'm now engaged to be married and, frankly, I'd like to manage it as soon as possible."

"Congratulations, Matt. Yes, you're right – it would be best if you could 'do the deed' within the next few weeks, before the pressure really builds up on us all. I'm sure we can squeeze out a week for a honeymoon if you move fast."

"Right, Sir. That's all I needed to know."

Matt's telephone call to Joyce was followed by urgent queries to his family, vicar, caterers, taxi firm and all the providers of wedding facilities. Brenda accepted happily to undertake to act as Matron of Honour, and also fulfilled her promise regarding the wedding dress, which was hastily adjusted for size. Matt's search for a best man was aimed at Reg, who was only too pleased to accept, although he didn't quite know what his duties would be.

Life on the unit was particularly active at this stage, as preparations began to be made for the major changes scheduled for the immediate future. It was rather like preparing for the Air Officer Commanding's inspection, only more so. Although they didn't actually have to whitewash the coal, as in the traditional Army joke, they did make efforts to clear away all the junk that had collected, clean up oil spills on hangar floors, etc.

The day of the wedding opened to a cloudy sky, but the Met Officer predicted sunny periods later. Matt and Reg managed to organise a lift to the coach station, and travelled up to London in a somewhat uncomfortable style, trying to avoid crumpling their best blues. A bus took them to Brockley where they joined the family. Joyce and Brenda were already there, as was Dennis, who had kindly escorted them both from Euston, having had to call in at Australia House regarding some domestic business. Brenda's marriage to Norman, a merchant navy officer, had gone smoothly, but he had had to return to duty as soon as their honeymoon was over.

Joyce gave Matt a warm kiss of welcome, then all three ladies disappeared into the bedroom to do what all ladies have to do to pretty up for a wedding. The men quickly freshened up and then, feeling somewhat 'de trop', and to relief all round, decided there was just time for a swift half at *The Jack*. There was no doubt that their absence enabled the preparations to be concluded that much sooner.

The whole company then departed, in relays, for the local church. Both men felt the need to ease their collars as they endured the seemingly interminable wait for the organist to strike up *Here comes the bride*. But eventually he did, and Matt permitted himself to turn and survey his bride proceeding sedately up the church on the arm of a proudly pink Dad. Her dress fitted her as if bespoke, and everyone agreed with Matt that she looked an absolute picture.

She joined him with her eyes glistening and her smile radiant. She was being united with her own man, and he had demonstrated his respect for her, and his deep love, so well.

To Matt, the next few minutes were somewhat hazy, but he recalled later that he had made his responses, when instructed, in a firm voice, and was so relieved that Reg produced the ring on cue without any embarrassing searching of pockets or under the pews, and that he, Matt, was able to slip it on Joyce's finger without dropping it – just some of the nightmare scenarios that had exercised his mind during previous uneasy nights. After the signing of the register, the assembled company retired home where a modest spread had been arranged, based on as many food coupons as could be collected from numerous willing donors. Conversation was lively and relaxed, with everyone enjoying the occasion.

Came the time for speeches. Following the sound of a tumbler receiving grievous bodily harm from a spoon, Reg rose to his feet, looking just a shade nervous, and delivered the following:
"Mr & Mrs Matt Gregg, Ladies and Gentlemen. I was very pleased when Matt asked me to 'do the honours' by standing behind him and seeing that he didn't chicken out at the last minute."
Slight laughter from the audience.
"On the contrary, I found I had to walk fast to keep up with him – he was that keen! You only have to look at his lovely bride to see why."
Slightly louder murmurs of approval. Reg was getting into his stride, and beginning to bond with his audience.
"Matt and I are what might be described roughly – and I use the term advisedly – as Aeroplane Bashers. This permits us to do things to aeroplanes that really should not be talked about in polite company. But we both like to think that we earn the pittance the Air Force grudgingly puts our way from time to time. Seriously, I must say that it is a pleasure to work with him, and I shall miss him during this next week. But I rather suspect that he won't miss me! I would now like to thank Brenda for acting as the Matron of Honour, and 'Mum', 'Dad' and the rest of the family for all they have done to help make this day so enjoyable for us all. And, of course, thank you, all of you, for being here. Now, please be upstanding while I propose the toast to – The Bride and Groom.

"I will now call on Dad to say a few words."

Dad cleared his throat, and rose to his feet clutching some notes in one hand, and adjusting his spectacles with the other.
"Thank you, Reg. As you all know, this has been a very dramatic year, particularly for Mum and me. First, the loss of our dear son, Mike. But now the gaining of a lovely daughter. As the saying goes, when one door closes, another opens. We have not had the pleasure of knowing Joyce for very long, but despite that short time, she has already entered our hearts, and we are delighted to make this family link between the Greys and the Greggs. We are so sorry that Mr Grey is no longer with us, but very pleased to welcome Mrs Grey and her son Alan into our family circle."

At this point, Dad turned the page of his notes, dabbed his brow with his handkerchief, and continued. "Matt has always been a source of pride to us, and we have become accustomed to his good judgement whenever it came to making decisions. His marriage to Joyce is clearly the best one he has made so far. Of course, in the future he will have help in making decisions. In fact, judging from my experience with Mum over many years, he may find some of these are made for him! It'll be a case of Choice or Joyce."

Smiles all round, and vigorous nodding by Mum. Joyce gave her a meaningful smile. Dad was clearly very pleased with his little joke.

"In all seriousness, we wish our happy couple all the best in the world. And my personal wish is that they have as much happiness with each other as Mum and I have had together. And now we must give Matt a chance to speak to us."

Dad sat down and took a long drink from his glass, contented but clearly relieved that his duty was now over.

Matt stood up sporting a smile of such width that it could only have edged its way out of the room with the aid of shoe-horns and the doors taken off. "Thank you Reg for your support. And thank you, too, Dad. Ladies and Gentlemen. There are so many things I want to say to you today that it's difficult to know where to start. You see, I've never done this sort of thing before!"

"I should hope not!" came the retort from somewhere in the audience.

"However, I can say, and I do say, that Joyce and I are delighted that so many of you could gather here and join us on our happy day. Unfortunately, my grandparents and aunts have not been able to be with us because of age and infirmities, but we have been assured that we have their blessings. Of course," he continued in a serious tone, "Mike would have enjoyed himself hugely on this occasion, and we all remember him with deep affection. I could never have thanked him enough for bringing Joyce and me together, and being the first member of our family to meet her – and not putting her off me by dragging out any items from my murky past. He was a Gregg Ambassador par excellence. I know that Joyce has already entered the hearts of all of us, as Dad has said, as well as my own. I would like to say right now that I am so grateful to my family for giving me life and bringing me up so that I could experience this wonderful day. And I am equally grateful to the Grey family for producing such a lovely daughter, and now entrusting her to me. You can all be assured that I shall take the greatest care of her for ever more."

Matt sat down, clearly emotional, and reached for his glass. Reg then rose and said: "Ladies and Gentlemen. We understand that Dennis has asked if he might say a few words. Dennis, over to you."

Dennis uncoiled his six foot two height to the vertical, and spoke as follows: "Ladies and Gents. I would just like to add my good wishes to Matt and his Sheila on this special day. I first met Matt on a course in Derby where we were sent to learn which end of the engine the bangs come from, and how to reach home when the tanks are nearly empty. As you know, I come from Australia, or 'Down Under' as you Poms call it, and I've often been asked, 'What's it like to be the right way up?' All I can say is, if this is what it's like, than it's pretty right by me.

"My birthplace is a lakeside township by the name of Woy Woy. I should explain that the Aboriginal name for water is Woy, so Woy Woy means Deep Water – it's as simple as that. When the folks knew I was coming here, they asked me to hunt up some distant relatives, but this proved a problem because nobody could give me directions to Sluff. It was only when somebody pointed out that Slough was pronounced to rhyme with Cow that I succeeded. Good job they didn't live in Loughborough. I'd still be looking!

"The item of news I really want to give you now is that I have been lucky enough to persuade Val to tame me into an engagement – and that 'Mum' and 'Dad' have both agreed. Sorry I couldn't contact you about this, Matt, but it's only just happened, and you must admit that you've been a bit preoccupied lately! Anyway, both Val and I hope you approve. Thank you."

Matt was on his feet even before Dennis had finished speaking, and he rushed over and pumped Dennis's hand with enormous enthusiasm. "Well, Dennis, aren't you the crafty one! There was I trying to teach you the finer points about English beer, but all the time you only had eyes for my sister. Of course we are delighted with your news, aren't we Joyce?"

"We certainly are. You know," she added thoughtfully, "I have the feeling that Valerie, Brenda and I should have some girl talk together."

"Too right," commented Val. "You see – I'm learning the Strine lingo already."

"Ready when you are," offered Brenda.

Came the moment for the bride to don her going-away outfit, followed by moist-eyed farewells from family and friends as Matt and Joyce took a taxi to a London terminus, and then a train to a quiet Dorset town. Military restrictions precluded a seaside honeymoon, so they settled for some English countryside instead. Matt checked his handkerchief carefully to see whether some joker had loaded it with confetti, but it was clear. At the hotel, they did their best to appear casual as if they had been long married, and seemed to be getting away with it until a middle-aged lady, meeting Joyce in the corridor, congratulated her on being just married to a handsome husband.

"But how on Earth did you know?" queried Joyce.

"Well, dear, at breakfast you asked him if he liked cornflakes. If you'd been married long, you would have known. Also, it was obvious from the way he kept looking at you. Do have a lovely marriage."

Chapter 8 – Postings (1943)

Matt returned to his Unit with mixed feelings – regret that his magical sojourn with Joyce had drawn to a close, for the moment anyway – but excitement at the prospect of new adventures on the horizon. He felt an added maturity with the loss of his virginity, but quietly proud that this had happened, in his view, in the most responsible way possible. He was aware that he, and Joyce, had much to learn about the techniques of making love, but he was content in the knowledge that they would learn these together, and excuse each other for any clumsiness in the process.

"Was everything OK?" queried Reg delicately.
"Just wonderful," replied Matt. "I'm now trying to get my mind back to work. What's happening?"
"Quite a lot. The Group bigwigs are coming in a couple of days to sort us all out for new jobs. The boss seems a bit cheesed off – he reckons he'll be sidelined to a desk job. You know how he hates all this paperwork, which is why he's always been glad to get us to do it."

The senior officers from Group HQ duly arrived, and interviewed each officer in turn. It was much as the boss had said, in that a surprising amount of attention was given to each officer's preference. Matt opted for an RSU and, after some deliberation, was called back into the office for Group's decision. This was to the effect that he would be posted to a soon-to-be-formed RSU to support a Wing comprising three squadrons of fighter aircraft engaged in bomber support and photo reconnaissance. It would be a Flight Lieutenant's posting as a Chief Technical Officer, serving directly under a Squadron Leader engineer as a CO.

These plans provided a perfect excuse for a series of farewell parties, and the whole station buzzed with excitement. Matt and Reg exchanged their news over foaming jugs. Every so often, the frustrations and nervous energies of the servicemen erupted in the form of a spontaneous party involving all sorts of horseplay. One of their favourite larks was to take the doors off the toilets, lean them against the tables, and attempt to ride over them on bicycles while being cheered by their colleagues. This was called 'Playing Aircraft Carriers'. Another game was to build a pyramid of tables, chairs, and anything else not screwed down, in order for one member to climb to the top, lie on his back and draw his footprint, with his signature, on the ceiling. Both exercises were undertaken that night, and the signs are probably still visible till this day.

"I see you're having S/Ldr Hammond as your CO, Matt. I've met him – he's a great guy, and boy, does he know his stuff. You're the lucky one. I'm also going to an RSU, but the word here is that our CO is a bit of a shocker."
"I certainly am lucky by the sound of it. I do so want to serve under a good boss. But first, I believe we have to go on a Battle Course at Baginton, near Coventry."
"That's right. You're due to go next week, and me the week after."

The following week saw Matt drawing a rifle and bayonet from armaments, and catching a train for Coventry. At the Battle Course centre, a number of hazards had been arranged in the form of an assault course, and the trainees were rapidly introduced to a daily routine of early rising, bayonet practice, breakfast (involving the demolition of a mountain of toast and a lake of tea), square bashing, more bayonet practice, lectures on survival, live ammunition firing, grenade throwing and evening meal, the rest of the day being their own when they queued for the ablutions block in order to indulge in healing hot baths. It was tough work, and Matt's hands and feet were becoming raw, particularly after he slipped from the swinging rope and landed in a trough of water. But, he had to admit during the return journey to his Unit, he had never felt so fit in his life. (He was later assessed as suited to being a leader in battle.)

Matt had apprised Reg by phone of what to expect on the course, and he then set about preparing his gear for a rapid departure when the call came. And come it did. He was to report the following day to an RAF station 'somewhere in England' where he was to start assembling the personnel of his new Unit. His CO was to follow within five days. When S/Ldr Chase gave Matt this news, he suggested that they have a farewell drink in the bar that evening. Matt was delighted to be invited, but somewhat saddened by the circumstances.

Conversation at dinner that evening centred on the imminent break-up of the Unit, and the directions in which the various members were to go. Matt would be one of the earlier ones to leave, so he would miss the actual demise of the OTU, but he was not sorry. Farewelling old friends was not a particularly attractive pastime as far as he was concerned – although he much regretted that he would be gone before Reg returned from Coventry. He would look for a chance to meet up when they settled into their new posts, or even when they eventually went into action in Europe.

Matt relaxed in the bar with a beer, wondering how he would cope working under a new CO at the RSU. S/Ldr Chase arrived looking a little harassed, and insisted on buying Matt another beer. He talked generally at first, but gradually warmed to his theme of reminiscing over the highlights of his service career.

"You know, people just don't realise that when I started in this business, we used to fly in biplanes with rotary engines – shaped like radials but with the crankshaft bolted to the airframe so that the whole engine went round with the propeller. Gave a weird centrifugal effect – when the aircraft turned one way it dived, and when the other way it climbed." That, thought Matt, reflecting on his university days, would be due to the Coriolis effect.

"Also," continued his boss, "it had to be lubricated with castor oil. It worked all right, but it streamed back into the pilot's face, and you can imagine the effect this had! Mind you, they were a mad lot in those days. There were a couple of brothers – both senior officers – who took off in a tandem cockpit aircraft with the older brother sitting in the front. When they landed, he was sitting in the rear – they had climbed out and changed places in mid-air! Of course, those early kites could almost fly themselves, but it was still a daft thing to do. But what characters they were, and no mistake.

"After you old boy"
"No after you old chap"

"Excuse me Sir!"

There was also a case where a rigger became airborne quite unwittingly. In the old days, of course, riggers and fitters would accompany pilots on their proving flight after repair, but you can't do that with single-seater jobs. You know how we get people to drape themselves over the tailplane to stop it lifting when running up the engine to do a mag drop test, well this particular chap didn't hear the pilot say 'OK, finished,' and suddenly found himself still clinging on as the plane took off. Fortunately, the pilot instantly recognised the problem and managed to keep his Spit stable as he carefully turned in to land. He was able to come to rest without losing his passenger, who then managed to ungrip the tailplane, stagger up to the pilot, salute, and say, 'Bloody good show, Sir,' and then collapse in a dead faint. What an experience! It was different with that enterprising pilot who took off in a Spit with his girlfriend sitting on his lap. He managed to fly without accident, but was spotted on landing and had a ginormous strip removed.

They also dealt pretty severely with trainees who transgressed with low flying, as they still do. It puts the frighteners on people on the ground when aircraft beat them up – one young trainee even let off his guns in apparent *joie de vivre*, not thinking there'd be any harm in aiming at a cemetery. What he didn't know until afterwards was that there were people there attending to the graves – they had to take cover behind the tombstones! And, of course, there were any number of cases where trainees tried to prove their mettle by flying under bridges – one even went through Tower Bridge, I believe. Utterly stupid, of course, but you can imagine the feeling of ecstasy when they first found that they could really fly an aeroplane." Matt wondered if Mike had ever got up to tricks like that – he wouldn't put it past him.

S/Ldr Chase continued, "One of the snags we had was when a fitter replaced a propeller on a Spit but didn't have time to complete the job before there was a scramble. The pilot presumed that his aircraft was ready, so he climbed in and started the engine… whereupon the propeller took off by itself and described a graceful arc across the runway, with the engine screaming its head off back at the aircraft.

"Yes, it's been a great experience, but it hasn't all been fun and games. There was a tragic case where a Spit was coming in to land at the moment when the wind changed direction. Unfortunately, the message didn't get through, and a second Spit started its take-off run from the other end at the same time. There was a bit of a rise in the middle of the runway so they couldn't see each other – difficult at the best of times with the tail wheel down. Now you know how wide runways usually are – plenty of room for two aircraft, you'd think, but curse the luck, these two Spits approached each other dead in line, and they went slap into each other. Immediate fire – ammo exploding all over the place. We tried to get the pilots out but didn't have a prayer. Makes you feel sick, doesn't it?"

"It certainly does," agreed Matt, as he thought back about the Whitley.

The boss was well lubricated by now, and embarked on the next chapter of his service life. "There was another shaky do we had when one of the trainee pilots was involved in formation practice. He idly touched the firing button on the stick, but he'd forgotten to switch it to Safe. Since his position was dead astern of the Chief Flying Instructor, there was hell to pay when they both landed – the student by plane, and the CFI by parachute! The lad was posted soon after that! I think he finished up out East somewhere, low flying over the desert in clipped-wing Spits. Pity they had to chop the wing tips off – that wing is a classic shape."

Matt thanked him for his reminiscences, and felt that it was about time he headed for bed. But the boss hadn't finished. Clearly he had something on his mind. "You know, Matt," he said, somewhat thickly, "I've done a very stupid thing in my life. I was posted to a remote airfield once, and met a girl at a local village hop. I hadn't been home for ages and felt very lonely, so when I took her home – well, the inevitable happened. I shouldn't have, of course, because I'm married, but Hell, you have to let off steam sometimes. Well, I thought it was all forgotten, but no way. It appeared that this girl had always been lonely and frustrated because of no boyfriends, so my attention wakened her emotions, and she's never left me alone since. Follows me everywhere I'm posted. God knows how it'll end. Perhaps an overseas posting is the only answer. But it's so unfair to the wife, and to the girl. Take my advice, Matt, and don't ever do anything like that – or even think about it."

Matt assured him that he understood the problem of loneliness when away from loved ones for any length of time, and that he would heed his advice.

He then tactfully drew the conversation to a close, and steered the boss towards his quarters.

Well, thought Matt as he lay in bed that night, who would have thought it? We learn something every day. He then kissed his photograph of Joyce, turned over and fell into a deep sleep.

Chapter 9 – Overlord (1944 – 45)

Transport arrived early the next morning and, after a hurried breakfast, Matt loaded his gear and glanced back nostalgically at the gate, having left a note for Reg in his pigeon hole. There was little conversation on the way, as Matt was preoccupied with his thoughts, and the driver with his route. Arriving at his destination, Matt was allocated his quarters, and then went to survey the hangar and offices set aside for his new Unit. Two aircraftmen turned up, one of them fortuitously being a stores clerk, so all three set to work sorting the items already allocated to their Unit and dumped in the hangar. Matt was somewhat amused to find that the largest item stored for them was an enormous grey wooden box containing dozens of toilet rolls. But then, this is how it should be, he realised. You don't just organise inputs of accommodation, clothing and food for your personnel, you have to consider the outputs as well!

Matt then visited the Flight Sergeant in the Motor Transport section to enquire about vehicles allocated to his RSU. "Yes, Sir. So far we have two Jeeps, one fifteen hundredweight truck and two 3-tonners here for you. We are expecting more lorries, two mobile cranes and some high- and low-loader artics within a few days. There'll also be some motorbikes coming your way. Business is brisk these days, and no mistake."
"Thanks, Flight. I'll send some drivers along as soon as they arrive."

The next couple of days saw the progressive arrival of Unit members, starting with F/O Reynolds (Salvage), F/O Dawson (MT), F/O Vickers (Stores), and F/O McKay (Repairs), together with an impressive array of aircraft fitters, riggers, drivers, clerks, cooks, batmen and a medical corporal. Then S/Ldr Douglas Hammond arrived, and Matt introduced himself and the other officers. The new CO greeted everyone affably but firmly. There was clearly going to be no nonsense with him around, and this RSU was destined to be one of the best under his leadership. This became apparent with the arrival of F/O Renshaw, the Adjutant, who expected that he would be handling all the incoming paperwork, and passing it along to the CO after perusal. But no. The CO ordered that *he* would receive it all and pass the relevant items back to the Adjutant and all other personnel involved. In this way, he would keep his finger on the pulse of *every* activity in which the Unit was involved. The Adj was not best pleased with this arrangement, but accepted it philosophically. In fact, he had no option.

Matt could now unload his initial responsibilities of general admin, stores and transport over to the officers concerned, and concentrate on building up a repair and salvage policy in discussion with the CO, McKay and Reynolds. The first thing was to arrange a suitable tool kit for each man, depending on his job, then to modify all vehicles as required. One thing they did was to bolt a vice onto the massive front bumper of a mobile crane, ready for instant use in the field. The need for practice in driving a wide range of vehicles, and in convoy, was stressed, and a small mound of earth was constructed, apparently of the same dimensions as the loading ramp for the Channel landing vessels, all personnel being required to practise negotiating it in readiness for their crossing.

A signal arrived from Group to the effect that all ranks were to be kitted out with battledress, reserving their best blue uniforms for formal occasions only. Matt smiled to himself, thinking, 'Well, I was wrong on that one. Never thought they'd do it.' Shortly afterwards, a second signal arrived to the effect that, since blue material was currently in short supply, khaki battledress would be issued in the interim.

The CO arranged to meet his opposite numbers in the three Wing squadrons his RSU was supporting, and he took his own officers with him. Useful contacts were made that would prove invaluable in the action to come. Matt was intrigued to meet an old school friend who had volunteered for aircrew following the loss of his younger brother in air combat.

Matt relished the technical efficiency with which his Unit was being created, and the speed with which their duties were being accomplished. There were snags, of course, but the CO declared that his door was always open, hence remedies were soon devised and implemented. It was demanding work, and Matt was always glad to hit the sack at the end of each day. Bed always seemed most inviting first thing in the morning when muscles were comfortably warm, but there was no opportunity for lying in at these times.

Came the morning when Alec Dawson from the MT office arrived in the Mess and announced that all vehicles would have to have a red, white and blue RAF roundel painted on their bonnets.
"What's the big idea, Alec?"
"It's to give immediate recognition from the air so that we don't get strafed by our own people. What's more, if you look outside you'll see that all the Wing aircraft are painted with broad black and white stripes at their wing roots. Same reason. Chaps, it won't be long now!"

And it wasn't. Two mornings later, on 6th June, the CO collected his officers together and announced: "Gentlemen, Allied Forces have just landed in Normandy. Operation Overlord is GO. We shall be moving any time now. Be prepared for anything, and good luck to us all."

But this move turned out not to be as imminent as they expected. The Wing and its RSU had been judged as particularly effective, so the powers-that-be had decided that they were capable of holding back for a few weeks and taking over the duties of their departing sister Units.

The complete Unit was then moved to a tented location on Gatwick racecourse where they prepared for embarkation. One of the worrying developments at that time was the German V1 flying bomb offensive. Although there was usually sufficient time to seek shelter between the end of the buzzing noise and the blast of the impact, some 'doodlebugs', either by accident or design, glided silently for some distance to their explosive end after the pulse-jet shut off. This was nerve-racking, so there was a rush to dig trenches within each tent in order to sleep in relative safety.

Eventually, the order came to be prepared to move out in convoy the following day. The CO left, with his opposite numbers, to cross the Channel by air, so Matt took full responsibility for the Unit. It was at this stage that the acute distress of two of his crew came to his notice. For whatever reasons, both men appeared to be on the verge of nervous breakdowns, so Matt despatched them to the medical section. Two replacement aircraftmen were hastily organised, and Group informed, as per instructions. Sets of the latter arrived in a number of documents including a double-sided sheet of paper entitled 'Invasion Without Tears' outlining the procedures for transferring the Unit from its present site to its planned operational site in Europe. This entailed going through what became known as 'The Sausage Machine' involving:

1. The Concentration Area, where vehicles are waterproofed, and a 4-day composite ration pack issued for use in the UK.

2. The Marshalling Area, where more waterproofing takes place, and a 2-day mess-tin ration pack issued for use in Europe only.

3. The Hards, for final waterproofing, and embarkation.

Two major documents had to be completed for handing in to the appropriate staff on the way, and there were various instructions about what to carry (haversack ration + matches to light tommy boiler + chalk for marking vehicles), what to do (keep engines clean + fit tow ropes at front and tied to top of cab), what was permitted (books, playing cards, footballs, cricket balls), and what was not (surplus kit, fishing tackle, grand pianos, pet dogs).

Everyone rose early, and the morning mist hanging over the racecourse was punctuated by bonfire flames disposing of all unwanted combustibles. The scene was somehow both ethereal and spooky. Two despatch riders arrived, and the convoy was assembled with Matt leading in the CO's utility car driven by Alec, the MT Officer. By opening the sunroof, Matt found that he could keep almost the whole of the convoy in good view with the aid of a pair of field glasses. One of the DRs travelled backwards and forwards along the convoy, reporting any glitches to Matt.

They duly went through the Sausage Machine without incident, arriving at The Hards in Southampton, where they recognised the hump in the loading ramp, looking identical to the earth mound they'd all been practising on for weeks. They started to load onto an LCVP (Loading Craft, Vehicle Personnel) vessel, which was equipped with a lifting platform to raise vehicles to the upper deck. Only one glitch spoilt their loading, when the driver of the 15cwt truck containing the telephone exchange missed his brake and hit the platform, slicing the radiator in two. Without hesitation, an identical truck was provided from a whole range of spare vehicles parked ready nearby, and the fitters spent the night transferring the exchange to it.

As the senior officer of the Unit, Matt was appointed OC troops, and when he walked round the decks the following morning he found that this LCVP was the leader of a whole line of ships, each one sporting its own barrage balloon at its stern. He wondered just how useful these would be as he recalled Dennis telling him that the practice in the RAF was to fit cable cutters on the leading edge of the wings of their bombers. If the cable got caught up with the propeller, there was not a lot one could do about it. But if it hit the wing, it would slide along towards the tip and eventually meet up with an explosively-operated cutter which would chop the cable in two. Very effective, apparently, if it met the cable, but one snag being that some unfortunate riggers had carelessly allowed their fingers to enter the cutter, so they were no longer – as Dennis put it – able to make a rude sign! Matt also remembered the order that no rings should be worn on the fingers in case they became snagged. A friend of his had jumped out of a cockpit on landing, and then realised that he had left behind a finger with its ring on when he went to pull the rope to handle the wheel chocks. Nasty!

The striking vision of the ships reaching back from the English coast made Matt regret that all ranks had been advised against carrying cameras for reasons of security in case of capture. What a sight it was, and what a record it would have been for future years!

Landing on Juno beach in Normandy went without any serious hitches. Although they had to drive through sea water, it was shallow, and the extension to their exhaust pipes, lashed up to the vehicle roofs, kept their engines running. They assembled on the beach near a road sign reading Beny sur Mer, and were told it would be some hours before they were to leave. There had been a hold up in the army's advance, and the city of Caen had only just fallen to the Allies. But even this eventuality had been catered for.

Each man then dived into his 24-hour rations and operated his Tommy Boiler. This comprised three small pieces of tin that could be interlocked to form a trivet with a small tray in its centre to accommodate a flat cake of paraffin wax shaped like a very short candle, plus a greyish block of a compressed mixture of tea, sugar and dried milk, a piece of chocolate and – oh, so thoughtful! – some sheets of toilet paper. They had brought their own water, and their mugs, so the Normandy beachhead rapidly hosted an extensive tea party, there having been no signs of the Luftwaffe at that stage. The CO joined them, and preparations were made to move off that evening. They all knew how savage the fighting had been during those first few weeks, and they were more than grateful that they had been spared so much.

Their convoy threaded its way between bomb craters through the French farmland, where several dead cows could be seen with their legs pointing stiffly into the air. Caen appeared to be one mass of rubble, with a road of sorts bulldozed through it and, as they proceeded inland through the countryside, Matt was intrigued to see several metal pipes running along the sides of the road. These, he was told, were carrying petrol to the army units ahead – it would be several years before he caught up with the supply end of PLUTO back on the Isle of Wight!

There was one episode that rather touched Matt's feelings. In all the villages they passed through, the inhabitants gave them an ecstatic welcome – cheering, clapping, smiling and offering them small gifts of fruit, etc. At one point, a farmer brought out a bucketful of fresh milk, which the thirsty 'lads' approached with their mugs at the ready. However, just as they were about to sample the welcoming fluid, they all paused, drew back and said in unison, "After you, Sir." Matt duly filled his mug and reflected wryly – either they hold me in respect, or they think the bloody stuff is poisoned! But he knew deep down that it was the former, as evidenced by the appearance one day of his likeness, over the letters 'CTO', painted on his mug by a budding artist in the Unit. You don't do things like that for somebody if you hate their guts.

At their operational site, the Unit was warned about the possibility of land mines, and to walk on established paths rather than open grass. Even that was not one hundred percent safe because some 'smart' mines would operate only when they had been stepped on several times. This did create some tension within all ranks, and proved to be the precursor of another case of a nervous breakdown later on. It was also evident that the Nazis adopted a policy of smashing and blocking all toilets as they retreated. This meant that sometimes the British forces, including Matt, had to straddle in open country across two tree branches arranged over a pit in order to perform one of the essential bodily functions. At least they had been provided with the necessary paperwork! The things I do for England! mused Matt.

All ranks were highly delighted when they came across propaganda leaflets left behind for them by the retreating Hun. One took the form of a series of cartoons showing a squad of Allied soldiers on the march, with soldiers progressively replaced by skeletons as each month passed, and asking if the effort was all worthwhile. Another gave reputed extracts from the American press to the effect that the Nazis would never endure the expression 'Unconditional Surrender'. But the one that drew the biggest laugh was the sketch of a British soldier in bed with a French girl, showing a picture of her French soldier boyfriend above the bed head. The hilarious aspect was that the Englishman was still smoking his pipe at the time! The lads all assured each other that they would knock their pipes out beforehand! One wonders just how the Nazi minds ticked because this attempt at demoralising the troops was enormously counter-productive by giving much enjoyed entertainment.

The inspection and retrieval of crashed aircraft went ahead much as it had been done back in England, the only differences being that they had to drive on the right-hand side of the roads – such as they were between bomb craters – and that they were occasionally shot at by die-hard snipers who had holed up after being left behind in the retreat, and were not minded to surrender. On one memorable occasion, the Unit experienced bullets buzzing through their campsite, apparently directed from a nearby belt of trees. Since Matt was Orderly Officer that day, he decided, somewhat inadvisably, to investigate. Keeping his revolver holster open ready for instant use, he moved quietly through the trees – to be confronted by a British Army Unit doing some ad hoc practice, but aiming a bit high over the earth bank! Keeping a tight hold on his temper, Matt pointed out the predicament, and would they mind aiming a bit lower in future, thank you very much! The Army instructors were most apologetic, and underlined their sincerity by presenting the Unit with a barrel of beer that evening.

The Poilu's revenge

One of the perks of being on active service abroad was that outgoing mail was free of stamp charges. Like many of the Unit's personnel, Matt wrote home to his wife every day, and this helped enormously in maintaining morale, and in the feeling of keeping bonds intact. One aspect Matt found distasteful was that all outgoing mail from the ranks had to be censored for security by the officers in case of interception by the enemy. Matt felt that it must be galling for a man to have his intimate letters, say, to his wife, subjected to the scrutiny of others, particularly those with whom he worked. Not that the contents of the letters were perused in any depth – in any case there was no time to do anything other than a cursory scan for sensitive items, but it was a chore that none of the officers relished. It's a pity, thought Matt, that we don't carry a padre on our strength. This would be just the job for him, and the men would probably prefer it, too. Matt was somewhat amused to find that a few members of the Unit took the opportunity to criticise their officers in print, in the sure knowledge that the words would be read by them. Matt didn't blame them for wanting to get their own back.

Other nuisance regulations promulgated as red tape from on high included the ban on wearing one's greatcoat collar turned up, as adopted by Hollywood heart-throbs trying to portray heroic military characters. Also, since aircraft frequently had to be worked on in the open, whatever the weather, the airmen perforce wore gumboots but these, it was stipulated, must not have their tops turned down to simulate equestrian or pirate boots! Someone in Whitehall must have been gleefully employed devising such nit-picking regulations. It was hoped that he would receive a suitable gong for his trouble.

A bombshell arrived in the shape of a letter from Joyce to tell Matt that her elder brother Alan had been killed in France. He was a paratrooper whose mission was to disrupt enemy communications but, after several successful drops, he had failed to return. It turned out that he had been shot just before making a landing. Matt poured as much sympathy as he could into his response, and only wished he could have returned to comfort her.

One other thing that did disquiet Matt to some extent was that, as the only Flight Lieutenant in the Unit, he was automatically saddled with the responsibility for the airmen's entertainment. In general, this caused him no concern, but he was rather taken aback to find that these responsibilities included the overseeing of the supply of condoms to those of the men who sought them. Not that he actually sold them – this was the duty of the medical corporal – but it struck Matt that the airmen's purchases might perhaps be guided to something a little more edifying.

Matt was not a prude, by any means. He had been through the workshops, and worked for a while in industry, so he knew the score, and enjoyed a joke, however risqué, as long as it wasn't downright filthy. He also was not averse to the occasional swearword if warranted, but there were certain words that he just would not use, and he never enjoyed hearing a woman swear. It was because he had such determined views about sex-only-with-love that he would have preferred to encourage others to adopt the same principle, rather than give the impression of facilitating the opposite. However, he accepted the task as part of his learning curve regarding leadership in service life, and was later to observe wryly that their condom funds became so extensive as to enable them to bid for, and purchase, a grand piano for the Unit when they eventually reached Brussels! Here again, thought Matt, this might perhaps have been a job for a padre… but perhaps not. Certainly not if he was Catholic!

The squadrons they supported had made quite an impression with the local French commercial community. By that time, four of the eight machine guns fitted to Spitfires, all focusing on a deadly point some yards ahead, had been replaced by two 20 mm cannons. The barrels were normally protected from rain, etc. by closed rubber tubes, which were shot off at the first burst of fire. When these tubes became unavailable, the armourers did no more than repair to the local *pharmacie* to buy up condoms by the hundred. This impressed the pharmacist greatly who was heard to pass a remark sounding like "Zeese Engleesh – *quels hommes!*"

Matt had already been intrigued to note that, as the engine power of the later marks of Spitfire was steadily increasing, and the Merlin engine itself graduating into its bigger brother, the Griffon, the number of blades on the propellers had to increase progressively from 3 to 4 to 5 in order to absorb it. He was also aware of contra-rotating propellers used in Coastal

Command but not, of course, of the 8-bladed propellers which were adopted many years later, although he had noted the experimental propeller comprising a single blade and a counter weight.

Joyce continued her regular replies to Matt's letters, giving him all the news from home. In one case she had to report the sad loss of Rex. The poor old boy had been taken for his usual walk in the woods, but barely made it back home, and clearly was so distressed that the family agreed that the Vet should put him to sleep. Well, he'd had a good life, and knew nothing but comfort and kindness, reflected Matt, but he regretted that he would never see the old lad again. Then came the news that Brenda had been terribly worried because Norman's ship had been torpedoed. He and several of his mates were in a lifeboat for some days until picked up by a following convoy. It seemed that the U-boats were not so numerous now, for which Norman was only too grateful. He had suffered a bit with mental stress, and was on sick leave.

The much better news was that Dennis and Val were to be married. The ceremony was to be held at a Registry Office in the presence of Mum, Dad, Joyce and Brenda, and they were all so sorry that Matt could not be there. Matt sat down that evening and wrote a letter.

> Dear Dad,
>
> I am so pleased to hear the news, and would be glad if you would read out this letter at the reception. Thanks.
>
> Dear Dennis and Val,
>
> I am thrilled to hear the news of your wedding, and only sorry that I can't be there to share the occasion with you. You have always been a good sister to me, Val, and a tower of strength to all of us in difficult times, so it is great to know that you have found such a good partner in your Dennis, and you, Dennis, in Val. You have my very best wishes for your future, and I trust you will find as much happiness as Joyce and I are finding despite our short times together.

Let's hope it won't be long before we all meet up for a really grand get-together.

Much love to you both.

Matt

Shortly after, the Unit was faced with a rather hairy situation. For some time, a steady stream of V1 flying bombs had passed overhead bound for London and other English targets, although their effectiveness was being steadily eroded by the integrated defences of fighter planes and Ack Ack fire, plus the winkling out and destruction of the launch sites, even though some of these were mobile. On this occasion, however, their Flight Test Pilot was returning after checking a repaired aircraft when he encountered an approaching V1 and, in his enthusiasm, shot it down. All very praiseworthy, of course, but he had momentarily forgotten that the RSU was directly underneath! Fortunately, the wretched thing glided on and crash-landed outside the Unit's campsite, but it cost the pilot several pints of beer afterwards to mollify his colleagues.

On the technical side, Matt was intrigued to learn that, as these buzz bombs flew at a speed of some 400 miles per hour, which the British fighters could not quite match, the more powerful Typhoon and Tempest aircraft fitted with the 24-cylinder sleeve-valve Napier Sabre engines (the Merlin and Griffon had 12 cylinders) were modified to tackle the job, and were fuelled up with a specially prepared aviation gasoline made with methyl aniline, water, methanol and other blending agents, plus additional TEL, to raise the anti-knock quality from 100/130 to 115/145, and higher. This permitted interception, however care was needed in tackling these weapons since it transpired later that one pilot who managed to home in and fire at one was so close that he was affected by the mid-air explosion and, in fact, subsequently died of internal injuries. A new attack strategy was devised as a result. One of the things that intrigued Matt was that the V1 was propelled by a relatively simple pulse jet, described as a 'flying stovepipe', apparently running on a fairly low-quality motor gasoline – an ordinary sort of auto petrol – yet it could fly at speeds that could only be matched by the best aircraft we had, powered by the best engines, and using the highest quality aviation gasoline that was made available for the purpose.

A letter arrived from the newly-wed Val (now Mrs Valerie Walker) to thank Matt for his good wishes.

> Dear Matt,
>
> Dennis and I thank you for your kind thoughts to us on our wedding day. It was a lovely occasion, and we look forward to all being together again soon. Although Mike can never be replaced, we felt that both Mum and Dad were comforted by the thought that the family had gained another son.
>
> I am glad to say that everyone is well here, and at Brickhill, although Gran is getting a little frail. It is so heartening to find that the raids are far fewer in London, so we can now get some peace at night. Also there is a new spirit about, as Dad confirms that the ARP arrangements are much more organised. As soon as there is enemy action, the defence people swing straight into action, with ambulances and, of course, our fire engines arriving on the spot within minutes. We seem to have come a long way since those terrible days of the Battle of Britain. I thought this news would comfort you.
>
> I see Joyce whenever I manage to get up to Brickhill, and we enjoy a good old natter together, as you will appreciate.
>
> Look after yourself, my dear, and here's hoping you'll all be back home soon.
>
> Love
>
> Val

The Unit received a sad blow when one of its salvage team drove over a mine and was killed. The medical orderly tried to go to his aid, but was ordered back because of the possible presence of other mines. However, the same neighbouring Army Unit rapidly arrived with mine detectors and cleared a path to the site. This was the Unit's first, and hopefully their last, fatality. Considering the conditions they had met throughout the campaign, they had been extremely lucky.

Quite early on, Matt had managed to find an electric windscreen wiper, which he salvaged from a wrecked car and fitted to his Jeep. After all, he argued, since he now belonged to the British Liberation Army (BLA), it wouldn't hurt to do a little liberating once in a while! In fact, he was somewhat astonished to find so many cars, many of them of high quality, lying about abandoned because the owners had completely run out of fuel. He'd had several offers to buy some of his service fuel (dyed red for identification), but daren't accept, despite any sympathetic feelings.

He'd also noticed how deathly cold a large continent could become in winter, far more noticeable than in the sea-surrounded UK. No wonder the Germans had adopted double glazing in their homes. This was something we should learn from them, thought Matt. He therefore found it expedient, when driving, to keep his hands down on the wheel at the 5 and 7 o'clock positions since this helped the blood to reach his fingers and avoid their becoming frostbitten. It was common practice to fit chains to the driving wheels in order to cope with the enduring ice. When working in the field with his airmen in such weather, it was Matt's habit to supply them with a tot of whisky each day. This was not as altruistic as it might appear, but was because Matt hated the stuff, although the airmen cared naught for the principle behind the act.

The following letter then arrived from Dad:

> Dear Matt,
>
> Trust you are well. I'm sorry to tell you that your Grandma passed away yesterday. She was a good age, and had had a full life, although she never really recovered from the loss of my

brother Harry on the Somme during the First World War. She was a good mother to all of us, and Father always thought the world of her, but her time had come. Joyce very kindly came over and laid her out, and the funeral is to be next week. Your Granddad is bearing up quite well, and keeping busy, as usual, which will help him cope.

Our house was damaged a bit recently when a V2 rocket destroyed a nearby house, but thank God nobody was hurt this time. The roof has been repaired temporarily, although one of the problems is to sort out all our papers and things which were damaged by the rain before they got the tarpaulin on.

By the way, we knew the invasion was imminent because the Woburn woods have been stacked with piles of ammunition. No guards or barbed wire or anything. Amazing. Hope they clear it all up when the show's over.

We look forward to your next leave.

Dad

Matt realised, of course, that such an event must be expected when people approach ninety, but he was saddened nonetheless, and sat down that evening to reflect on those care-free days of childhood when a visit from the grandparents was always a cause of great excitement. The major item of interest was always "Granny's Bag" which was viewed as a veritable cornucopia of magic properties from which Granny would invariably extract numerous gifts of sweetmeats, coloured pencils, drawing books and the like, bringing shouts of glee from all three children. She was a kindly, motherly soul, quite tall in her younger years before the onset of osteoporosis, and with twinkly blue eyes. She had been 'in service to the gentry' as some sort of maid, and it was rumoured that she had been employed at Blenheim, serving the Churchill family… and had bathed the young Winston!

* * * * *

With the progressive advance across the Continent, the Unit eventually reached its semi-permanent site at one of the airfields near Brussels. This gave the men a chance to spread out a bit and find more space for their various activities. On the way, they had passed Group HQ which led to a certain amount of embarrassment. The Unit had made a practice of adapting captured German trailers for its own purposes. In one case, a lengthy trailer had been converted for use as a mobile cookhouse, the floor being strengthened with steel sheets. As luck would have it, the sheets decided to disintegrate and bring the trailer to a halt precisely outside Group's main office. Rapid repairs were executed, but there was no surprise when, on reaching their destination, a signal awaited them to the effect that no German trailers were to be used in future!

It was from this site that one of the fitters was granted compassionate leave to visit a sick relative and, on his return, spoke convincingly of seeing an aircraft flying *without a propeller!* Of course, he was told to stop talking out of the wrong orifice, but was insistent that he had been perfectly sober at the time. It still came as something of a surprise, however, when a signal arrived to the effect that this RSU was to take responsibility for repair and support of the new Meteor jet aircraft that were arriving for purposes of photo reconnaissance work. Of course, when they arrived, *all* members of the Unit found time to examine these unusual craft from as many angles as possible.

The booklets provided to the CO and his team showed that the jet engines fitted went under the names of Derwent and Eland. It also transpired that, being of early design, some faults might be expected in relation to their bearings. The check involved looking up into the jet pipe to see if any sections of bearing were stuck on the inner surface – an event that was happily quite rare. Needless to say, Matt was particularly pleased to be introduced to the new method of propulsion at such an early stage in its development. Coupled with the experience of the V1 aircraft, the end of the high-performance main aero piston engine was in sight!

It was not long before the rest of the world realised that the jet engine was here to stay, but that its immediate adoption was hampered by shortages in the supply of its kerosine fuel, with so much effort having been put into the improvement of gasoline – as in the defence against the V1. A decision was thus taken to ease the supply position by blending a so-called wide-cut fuel comprising the components of both gasoline and kerosine. Jet engined aircraft were subsequently cleared for this fuel in addition to the standard aviation kerosine, although it was presumed to be for a limited period only. But, as with pre-fab houses, this temporary measure was to persist for many years.

Matt was pleased to note the continuing progress in countering the V1 attacks. However, a new and deadly menace now arose in the shape of the V2 rocket, which dampened the rising optimism of the hapless civilian public. These fiendish missiles appeared to approach their target at so high a speed that no warning was available, or counter action possible. Destruction of the launch sites seemed the most effective answer, together with bombing the manufacturing bases, if they could be found, although this caused danger to the slave labourers employed. Matt had heard this trick of presenting a human shield of innocents, as was also practised by the Japanese when they were evicted from the Pacific islands. Matt resolved to learn everything he could about these weapons as soon as the details were forthcoming.

A heart-rending letter arrived from Joyce to the effect that Dennis's aircraft had been shot down, and he was missing! Matt replied at once, and did everything he could think of to make contact direct, but before he could organise this, a second letter arrived, this time from Valerie, to say that Dennis had been picked up by AirSea Rescue after several hours in the Channel, and was recuperating in hospital, having suffered some burns. Matt could almost taste the relief he felt, and was glad to be able to settle down to work again with a quieter mind.

From the news permeating the whole Unit, it was clear that advances into Germany itself were such that the war was at last drawing to a close. However, the Luftwaffe made one last death throw by mounting night attacks on as many Allied airfields as it could. Matt was driving round his airfield at the time that a German fighter/bomber screamed overhead – he decided to park his vehicle in the shadows until the danger was past. On 8th May 1945 came the moment they had all been waiting for… VE Day.

The Unit did what it could to mark the occasion by mounting a party in one of the hangars with copious beer supplies and some unfamiliar sweetmeats provided by the cooks, but there was some regret that they could not join in the jollifications that were apparently going on in Trafalgar Square back home. However, the local populace was equally overjoyed, and invited them to join their bonfire party in the city. This invitation was promptly accepted, and a good time was had by all – until the moment arrived when a small bomb, thrown on the bonfire by some idiot, decided to explode. One of the Unit's men was injured in the cheek by a piece of shrapnel, and had to be sent to a military hospital. Matt made a point of collecting him when he was discharged, and returning him to the Unit.

Chapter 10 – EVT (1945 – 46)

The transition from war to peace was both sudden and confusing. Although the activities of all concerned had varied widely throughout the campaign, their goal was common – to end the war by beating the enemy. That accomplished, the cement that had bound them all together was removed at a stroke, and many units, and their personnel, had to search for new objectives.

One thing that rapidly became obvious was that there would be no triumphant victory march through London for most of those serving abroad – although such a march was organised eventually. On the contrary, an extended period of 'sweating it out where you are' was the order of the day, with no indication of a date for permanent return home. This created a depression in changing from the previous excitement and danger to boredom and frustration, and the military leadership attempted to encourage projects of a constructive nature, although this sometimes meant no more than refurbishing a caravan for a CO.

Fortunately, ENSA continued to provide entertainment, as they had so nobly done throughout the campaign, and the units themselves promoted some impromptu concerts, with numerous 'in' jokes about their colleagues and predicament. Regular periods of home leave gave an obvious safety valve, but the major objectives were seen to prepare volunteer reservists and conscripts for return to civilian life.

The sinister effect of this unsettled post-war environment was brought home violently to the Unit when F/O Vickers, the equipment officer, was found dead in bed with a self-inflicted shot in his brain. God knows what his problem was, was the universal thought. If only he'd mentioned it someone might have been able to do something. It then transpired that this was not an isolated tragedy, as a number of people found the situation to be so unreal that they could no longer cope with living. Such a pity to have come successfully through years of danger and then to end it all in this way. Matt didn't envy the CO, having to write to his family about this. It was reported later that, during this confused aftermath, about 40,000 French collaborators were killed by their compatriots.

The longed-for VJ day arrived in August 1945, but only after the dropping of two atom bombs on Japanese cities. While rejoicing in the overall relief at the arrival of peace after both terrible wars, Matt couldn't help feeling that it was a pity that so many civilians had been killed by this action, even though many thousands of lives, on both sides, had no doubt been saved as a result. Matt wondered why the bombs couldn't have been dropped in open country to serve as a warning about what would happen if there was no surrender. Matt also wondered idly what would have happened if such a bomb had been dropped on Fujiyama itself – perhaps a gigantic earthquake!

Plans swung into action to bring home the survivors of the POW camps. Matt joined in the general revulsion felt at the state of the prisoners resulting from the brutal conditions they had had to endure, and when he heard the story of the Australian nurses who had been ordered to stand in the surf and were then machine-gunned, he vowed he would never forget or forgive.

Since the Allies had sent into Germany teams of expert scientists and engineers to ferret out the technical secrets behind the Nazi war effort, interesting information was coming to light about jet propulsion and rocketry. Matt was gratified to get his hands on the following details of the V2 rocket vehicle:

Dimensions	Performance
Length: 46 feet	Maximum thrust: 25 tons
Diameter: 5 feet 5.3 inches	Burn time: 65 seconds
Weight: 12.5 tons, of which warhead = 1 ton	Flight time: 5 minutes
	Average range: 185 miles
	Maximum altitude: 56 miles
	Final velocity at target: 1800 mph!

One of the items that intrigued Matt was that directional control was achieved by the use of four graphite rudders fitted internally within the exhaust stream to assist the external control surfaces which were ineffective at the low speed of take-off. The rocket motor was powered by 75% ethyl alcohol as the fuel, and liquid oxygen as the oxidant. These two fluids were transferred to the rocket engine by means of pumps energised by the steam and oxygen generated from a source of liquid hydrogen peroxide, the process catalysed by means of calcium permanganate. Although Matt marvelled at the technical complexity of the vehicle, he grieved for those who had lost their lives through its use, and hoped fervently that the knowledge gained would be put to good use in future for peaceful exploration of space.

In occupied Europe, a scheme of Educational and Vocational Training (EVT) was instituted to give instruction in such areas as Mechanical and Electrical Engineering, Building, Printing and Farming. Naturally, Matt favoured this scheme, and was delighted when his application to participate was accepted, and he was appointed as Head of the Mechanical Engineering School. An abandoned garage had been located near one of the occupied German towns, and Matt was instructed to scour the countryside for any suitable machine tools that could be found in bombed-out factories and the like.

Matt and his team spent many days touring through the local mountain area and towns, backed by heavy lifting gear from their base whenever they struck lucky. In this way they built up a sizeable collection of screwcutting centre lathes, milling machines, pillar drills, shapers and power saws, plus a sheet metal bending machine and guillotine, a forge and anvil, drawing boards, a small wind tunnel and a smoke tunnel. These machine tools were installed on concrete beds made by the Building School, the power lines laid in channels by the Electrical School, whereas the ragbolts, brackets and belt guards were made by themselves. Also located was a valuable collection of sectioned aero engines, working engine models and technical transparencies which were added to their collection of visual aids.

Matt and his two sergeant assistants then devised a lecture course to cover the basic sciences, engineering drawing and workshop technology, plus a test piece involving turning, screwcutting, milling, drilling and knurling, which the students were required to draw and then manufacture using their own drawings.

On one early foray for machine tools, they passed by the infamous Belsen concentration camp, and were sickened to the stomach by the sight of the poor wretches awaiting transfer to hospital. They could understand such atrocities, perhaps, from some primitive tribe of savages, but the fact that the perpetrators were from a civilised and apparently cultured society seemed unbelievable.

In these early stages of the occupation, the average Germans appeared to be anxious to make friendly contact with their captors, but the Allied leaders initially frowned on this, and a non-fraternisation policy was imposed on all personnel. However, human nature being what it is, this ban was bound to be short-lived, and Matt was amused to note that the lads were being taken into town in one of their 3-tonners which had chalked on its side in large letters: 'THE FRAT WAGON'.

It had been noticeable throughout the campaign that continental families wanted to make close contact with the Allied troops, particularly the British. This was partly due to the need for food – and payments were offered for any items that the troops could scrounge from the cookhouses – but, more fundamentally, due to the desire to see their daughters married off to British husbands. To this end, the bizarre situation arose in which parents would openly offer their daughters, complete with bedrooms, to likely spouses. This, of course, was diametrically opposite to the experience of many a virile young serviceman at home who had to risk the wrath of a father in any attempt to seduce a daughter. In fact, this situation had been encapsulated in one of the less ribald songs routinely rendered at beery Mess parties, somewhat on the following lines:

>Behind the door, her father kept a shotgun.
>He kept it in the springtime, and in the month of May;
>And if you wonder why the hell he kept it,
>He kept if for an airman who is far, far away.

This situation became so rife that instructions were issued from Group that airmen should be discouraged from forming permanent relationships with local families. Matt tried his best to comply, but how easy is it to change the mind of some lusty young man intent on following his instincts?

Meanwhile the EVT scheme developed steadily, with intakes of groups of 20 or so servicemen to learn the arts and crafts of 'metal struggling'. Matt found invaluable the support of his two sergeants, both of whom had extensive workshop experience. He also created good working friendships with his opposite numbers in the electrical, building, printing and farm schools. This was undoubtedly giving him valuable experience for a future academic career, even though it was workshop-orientated, but he was becoming increasingly impatient to start his post-war career proper.

One of the many impressions that Matt gained during this period was that the German womenfolk were noticeably wary when they enquired at the camp for work, but desperation for jobs, wages and hence food forced them to apply. In conversation with them later, it transpired that they had had unpleasant experiences of molestation and even rape from some Eastern servicemen, but that they had far more confidence in the behaviour of the British. (This was decades before the lager louts and football hooligans fouled the picture of British behaviour abroad.) When they found that they were treated with respect, and even given a room of their own where they could take their breaks, their gratitude was almost immeasurable.

Again there was tragedy when a young German soldier, brainwashed with dire threats about what the Allies would do to him if captured, could stand things no longer, and one night switched on the power to the overhead gantry, then climbed up a ladder and touched the live cables. His remains were found on the concrete floor the next morning. Fortunately, there were few instances of such events, and relations between Allies and German nationals steadily thawed.

Once again, Matt received bad news from home, via Dad:

Dear Matt,

You won't be surprised to learn that Granddad has just passed away. He seemed reasonably well during the last few weeks, but did not speak much and we feel that he just lost heart after losing his companion for all those years. The local undertakers are back in business again, and are attending to everything for us.

As you know, we now have a Labour government. How on earth can people discard Churchill like that? He saw us through this terrible war, and was the only man who could have done it. There's gratitude for you!

We hope you will manage to be demobbed before very long.

Cheers

Dad

Matt sat and reflected on the part that Granddad had played in those childhood years. Granddad had been a coachman to a wealthy family, and was meticulous in the way that he handled and cared for his horses, and in his own neat appearance. So much so that when his son, Dad, became involved with the emerging motor car, and learnt to drive ambulances, he could only look on these 'horseless carriages' with contempt, and cling ever closer to his way of life. Although of slight build, he was wiry with not an ounce of fat on him. In fact, he took a major role in the housework at Battersea, relieving Grandma of many of the more tedious chores. He was devoted to her and his family and, when their son was lost on the Somme he had put his age back to enlist in the Royal Marines. The fact that he was seasick every time he left the shore did nothing to diminish his courage or determination. His passing so soon after Grandma's therefore came as no great surprise.

As regards the new government, Matt shared his Dad's view that it was a pretty shabby trick to throw out a man who had done so much. But that's politics for you. What incensed Matt was to read that someone had described Churchill as a warmonger! What rubbish, he thought – here we were facing possible defeat with God knows what consequences of destruction and killings, yet Churchill had told us the truth that we faced blood, sweat and tears etc, and had piloted us through to victory. Of course he would have his 'warts' like everyone else, but surely those who disagreed with him would have the decency to recognise his success. But apparently not. For the first time, Matt realised that politics bore no resemblance to Applied Science where people search for the most probable truth, and then attack it to see if it holds up to criticism and, if not, dig further for the next likely truth. In politics, it seemed to him, people just said what suited their party line, and to hell with the truth. Politicians, Matt decided at that moment, were not people to be trusted or admired, whereas Churchill, as a statesman, was.

One of the items discovered during the machine tool search, much to Matt's delight, was a discarded CFR variable-compression engine, complete with bouncing-pin knockmeter. This was brought back to the school in triumph, and Matt then negotiated with the building school for the construction of a rugged base for the installation. The engine could never be refurbished for anti-knock rating of fuels, of course, because this is a delicate operation and needs skilled handling, but it could prove itself useful as a tool for exploring the effects of compression ratio on the combustion of any fuels of interest.

By now, Matt had found time to prepare a brief booklet recording the facilities and courses of his school, complete with some photographs taken by one of his staff. He then wrote a rather longer booklet incorporating much of the technical details of petroleum fuel production and use that he had learnt at university and since, and he managed to persuade the Printing School to prepare a coloured frontispiece to set it off. This booklet was to prove useful to him in future job interviews.

One day, they noticed that two V1 flying bombs, suitably disarmed, were being displayed in the square of the local town. This gave everyone an excellent opportunity to inspect the machines, and particular interest was shown in the one that had been fitted with a cockpit and flying controls. Matt managed to recall dimly that a problem had arisen with the earlier versions of the aircraft, and that a famous German woman pilot – he thought her name was Hannah Reish, but couldn't be sure – had volunteered to fly the thing and sort the problem out. Apparently, she was only too successful. Matt also seemed to remember that this gutsy lady had managed to fly a helicopter inside a hangar! What a pity she wasn't on our side!

Another source of pleasure for Matt was a visit from Reg, who had heard of the exploits of the EVT schools through the Group's newssheets. With his detailed practical background, Reg was able to make several useful suggestions to improve the course, and to design suitable test exercises to help the students develop their skills with machine tooling.

Inevitably, numerous requests came their way to manufacture toys, model steam engines, and the like, but Matt kept a firm hold on this since the first priority was the provision of suitable machining exercises for the students rather than the usefulness, commercial or otherwise, of the products. One request that Matt did accept was the turning of a pair of candlesticks for the local padre. A number of brass ingots had been collected with the machine tools, and these served admirably as lathe-turning projects involving tapered and curved surfaces.

When Matt discovered that an early class B release scheme had been introduced for teachers, who apparently were in very short supply back home, he prepared a letter of application and sent copies off to as many universities and colleges as he could think of. When one insensitive politician suggested that the letters BLA, instead of representing British Liberation Army, could be interpreted as Burma Looms Ahead, Matt's impatience increased even further. He had enormous respect for all the lads, and lassies, involved in the war with Japan, as he had for so many other people in the sister services, merchant navy, bomb disposal groups, fire and ambulance services, medical service, land army and just about everyone else involved, very much including all the womenfolk left at home – particularly mums with babies – but he felt he had done his bit for the war effort, and it was about time he started contributing to the nation's welfare via the academic route. He was also longing to return to married life, and to contemplate the start of a family.

Two things then happened simultaneously that put him in something of a spot. First, a letter arrived from a London college offering him the position (subject to interview) of an Assistant Lecturer. Second, a senior officer from Group approached him with the offer of a Squadron Leader posting to take over the complete local EVT scheme of mechanical, electrical, building and farm schools. An added attraction was that it appeared possible for Joyce to join him in Germany.

Matt went for a long walk, deep in thought about the two options. When he returned, he still hadn't reached a decision, but he had been able to clarify his thinking about the various implications, so he sat down and wrote them out:

1. A squadron-leader appointment is the first of the senior officer ranks, and is often used subsequently in civilian life.

2. The pay of a squadron leader is quite attractive.

3. He could be joined by Joyce and live a reasonably normal married life.

However:

4. He had no wish to continue use of the squadron leader title. Once he left the RAF that would be it. The only contact he might contemplate would be to join the RAF Association.

5. He would have to start at the bottom rung of the academic ladder some time, despite the pay, and remain there until he was brought up to speed with his university subjects.

6. He was not convinced that Joyce would enjoy service life despite, or because of, the many formal dinners, dances and other functions that this would entail.

7. He wanted to be reasonably near his parents and Joyce's family so that he could contribute practically to their welfare.

All in all, he concluded that he should bite the bullet and seek an interview at the college, then, if successful, apply for a Class B release. He put all these thoughts in a phone call to Joyce, and was much relieved to find that she generally agreed with him. She, too, felt the desire to start a family, and would prefer to be in England, close to relatives, when that happened.

Matt was due shortly for leave, and he found it wonderful to be back with Joyce and his family. At the college, the interviewing panel expressed their interest in his application. Matt thought they must be desperate for staff and were clutching at straws – even his. They did voice concern that he had not read academic literature very much since leaving university. Who the hell has – and who could? Matt thought, but they assured him that they would give his application every attention, and communicate with him very soon. They were as good as their word, and shortly after his return to Germany, Matt received confirmation that the job was his for the taking.

Matt accepted immediately by letter, and advised the EVT people accordingly. They regretted that he was unable to accept their offer, but understood his viewpoint. He then set the Class B release wheels in motion, and prepared to pack. He didn't want to leave the EVT scheme in the lurch, but suddenly had a brainwave and wondered if Reg might be interested in taking over the Mechanical School. It turned out that he was, and was so appointed, which enabled Matt to leave with a clear conscience.

Matt had 'liberated' a stout wooden German crate which he filled with most of his belongings, and screwed down leaving the padlock inside, as instructed, so that the contents could be inspected for dutiable items. It became clear that the authorities were very keen on intercepting guns of any sort, which some servicemen wanted to keep as souvenirs. Matt had included a rather nice magnifying glass amongst his gear, which he thought would suit Dad when reading the newspaper... but it was missing by the time his crate arrived home.

The long journey to Boulogne by train was much superior to the 3-tonner transport he had had to endure while travelling to and from periods of leave, and he was intrigued when he found himself billeted for one night in a nun's cell in a Belgian convent. Not that the nun was there, of course, but Matt could almost hear Mike saying, "Wouldn't have done you much bloody good if she had been!" Matt mused, 'Dear old Mike. What a joy he had proved to everyone who knew him during his short life. I wonder how he would have settled down in a post-war world? And what a wonderful uncle he would have been to any of our children.'

It was a curiously nostalgic crossing by ferry to Folkestone, reviving memories of holidays there with the family and Mike as a boy, then his more recent departure from leave as a man. He was, of course, delighted to be back, and thrilled at the sight of 'the White Cliffs', but somewhat saddened at leaving his friends. The Southern Railway took him to Waterloo, and he was glad to see that bomb damage was beginning to receive attention. He was given the option of a demob suit or a set of sports clothes to take home, and then boarded a bus to Brockley where he strode up the street, realising that this was probably the last time that he would wear his uniform. He was carrying on his chest a ribbon of the 1939-45 star, and an oak leaf awarded for being mentioned in dispatches, but was not over-impressed with either. He'd had a job to do, and he did it, just like everyone else. Matt was also somewhat jaded by the fact that some gongs seemed to be dished out as matter of routine – coming up with the rations, so to speak.

Matt was right about his uniform. After embracing him, Joyce eased him out of his best blues, and he never saw them again. What Joyce did with them he never knew, and didn't care to ask. That part of his life was over for good. His greatcoat, however, was a different matter. This was made of good quality Crombie material, and Joyce managed to remove the buttons and epaulets ready for civilian wear in winter. One of his friends remarked later, "Matt, do you think that style will ever come back?" To this day, he could never decide on a suitably witty response.

Matt now had the great joy of being with his precious Joyce continuously, and experiencing the many aspects of married life. Her companionship by day was something that, he realised with emotion, he had dearly missed while away. And at night, he discovered a magical way of cuddling her, whether or not they made love, by bringing her back to him as he lay on his left side, and placing his arm around her so that her left breast lay in his hand.

Although he knew that both breasts were equally beautiful, the left was the one he had kissed when they first became intimate.

Great news arrived from Brenda regarding their firstborn, Dulcie, who had put in an appearance almost exactly on the day predicted. Norman had left the merchant navy, and was taking examinations for the civil service.

By now, Dennis was out of hospital, suitably patched up and, although not yet demobbed, was not required to fly again. In any event, there was a move to replace flight engineers with automatic control instrumentation.

It was a joyous time for all of them to assemble for that long-promised meeting at Brickhill, with no enemy action or travel restrictions of any kind, and to enjoy each other's company in such peaceful and stress-free conditions. It came as no surprise soon after to receive the happy news that Val was now pregnant.

A poignant moment arose with the arrival of a letter addressed to Mike, as follows:

> Dear Mike, Boyo,
>
> How are you? I hope you are well. Sorry I missed you but I had a shaky do a few months ago, but they got me out before the fire was too bad. Glad to say that they patched me up with plastic surgery and I am now about as handsome as I was before. Ha. Ha. The best news is that my nurse was so good to me that I married her, and we live in her home in Wales. Also we are expecting a baby in the Summer – not everything was burnt! Ha. Ha. If it's a boy we are going to call him Mike, so I'll always have a Mike to speak to wherever you are.

Hope your luck is holding out.

Write a line when you can.

Regards

Teo

Matt had the melancholy task of putting Teo in the picture, before wishing him and his forthcoming family well. He then fixed Mike's Ozzy firmly above his desk in his small study.

Chapter 11 – College (1947 – 50)

Matt presented himself for work, feeling quite out of place when travelling in his best suit surrounded by dozens of other commuters, and could not face walking across the grass in the college grounds in view of the inbuilt fear of land mines still impressed in his mind. He also had to force himself not to salute when he was shown into the Principal's office. After a short pep talk of welcome, the Principal took him to see the Head of the Engineering Department who briefed him on the layout of the building, plus the structure of the degree courses and the part that he, Matt, was expected to play in them. It appeared that lectures were grouped into two periods of morning and afternoon, plus a third in the evening for the part-time students. The total contact time for a lecturer was stated as 30 hours per week, the rest of the time being reserved for lecture preparation and exercise marking, and Matt hoped fervently that his total would be far less to start with so that he could get plenty of hours in for lecture preparation.

Matt would have preferred to concentrate on thermodynamics, but there appeared to be a pressing need for lectures also in applied mechanics, materials and structures, theory of machines and mechanics of fluids, as well as workshop technology, so Matt was hard pushed from day one. One of his new colleagues, who had been there throughout the war, had lectured on just about every subject in the book, and was kind enough to loan him his notes and, equally invaluable, a set of worked examples that could be used as exercises and/or exam questions. Another colleague, ex-World War I, did the same with regard to thermodynamics, and Matt was more grateful than he could say.

One of Matt's duties was to reorganise the instruction sheets for the laboratory exercises, and he found himself handling once more a Fletcher Trolley, a simple piece of equipment comprising a weighted block of wood sliding along an inclined wooden surface under the action of a weight on the end of a piece of string, the object being to measure the coefficient of friction between the two wooden surfaces. As Matt tackled this, he thought back to the sophisticated aircraft and engines with which he had so recently been concerned, and how here he was reduced to playing with two pieces of wood! Momentarily, he wondered whether he had taken a retrograde step in his career, but quickly reassured himself that this had been, after all, the wiser move.

As regards his lectures, Matt had the confidence to handle these in the full knowledge that he had the personality to master the techniques of delivery, even though he had mugged up the material only the evening before, and he found it a great help that some of the students were servicemen who had volunteered before finishing their degree studies, and were being allowed back to complete. These people were on the same wavelength as Matt, with a shared sense of humour. It reminded Matt of how lucky he had been to have a wise father to advise him to continue his studies rather than join the Forces when the war started. "Matt, my boy," Dad had said, "Your country will need qualified people now so you will be more valuable if you carry on." Then a year later when France fell, Matt again suggested volunteering. "But Matt," said Dad, "Graduates will be needed even more now, so keep going."

Nevertheless, Matt felt eventually that, as his duties had crept up to over 29 hours out of the stipulated maximum of 30, this was unfair treatment for a newcomer after an unavoidable break of several years from study. When his complaint was received with no sympathy whatsoever, Matt decided that, after a suitable period of service to recompense their acceptance of him on Class B release, he would seek his fortunes elsewhere. It wasn't the level of the workload that worried him so much as the lack of understanding and reasonable compassion.

* * * * * *

Great excitement followed the announcement of the forthcoming royal wedding of Elizabeth and Philip, and Matt and Joyce joined in the rush to buy a 'televisor', as TV sets were called in those days. It had only a 9-inch screen, and was in black and white, but it seemed like magic to be able to witness all the news items and entertainment on offer.

By this time, Heathrow Airport had opened, reminding Matt how he and Mike had spent many a weekend cycling to the several airfields in and around London, taking photographs of the Handley Page airliners at Croydon, and of the Hawker biplanes at Biggin Hill. What blissful carefree days they were for the two of them!

On the domestic front, the great news was when Dennis rang excitedly to report that their baby had arrived, and that his name was to be Harry. Both Val and their offspring were chipper.

Life became much more interesting and exciting, though no less busy, when Joyce confronted Matt one evening and revealed that she too was now expecting. They were both overjoyed, as was the rest of the family. Since this news coincided with Dad's retirement from the London Ambulance Service, he and Mum decided that they would move to *High Firs*, and hand over the running of the Brockley property to Matt and Joyce. Dad had managed to buy himself a small car, but Matt had not yet had time to refurbish, or dispose of, Mike's old faithful Mabel still stored in the garage at Brickhill. Some of the furniture, scarred here and there with bomb damage, was carted from Brockley to Brickhill by hiring the greengrocer's van for a weekend, then Matt and Joyce settled down to prepare the house for their new arrival.

One essential for Matt was the setting up of an effective workshop in the garden shed so that he could carry out household repairs, and hopefully make a few toys and things for the infant. So one evening, when he was free of lectures, he called in to the tool shop after leaving college, and selected a number of useful items including a fairly heavy bench vice. He reckoned that he could just about carry all of them to the bus stop, and then home. But he chose the wrong day, for with little warning, one of those pea-souper fogs (later called smogs) that used to typify London – according to Hollywood films, anyway – rolled in and enveloped everything in a cold, clammy blackout.

However, it seemed that the wartime spirit was not dead, and the conductor, carrying a torch of burning branches, walked in front of the bus so that the driver could at least make some progress without colliding with any parked vehicles or mounting the pavement. This arrangement worked reasonably well, if a trifle slowly – although some swine of a passenger stole a packet of loose change left by the conductor under the stairs – but it meant that Matt had to debus about a mile from his normal dropping-off point. 'I would pick today,' he muttered, as he struggled along with his heavy load, and was dead tired on reaching home, with his arms trembling and feeling twice as long as normal. 'Ah well,' he concluded philosophically, 'perhaps it will be easier now to tie my laces!' Joyce had been worried, of course, but this was decades before the advent of mobile phones, and there had been no telephone kiosk en route that he could have used to set her mind at rest.

Not even that would have helped, because there was no telephone at home! It transpired that a significant number of lives had been lost – the fog damaging lungs and hearts – and this heralded the beginnings of the clean air policy, comprising smoke-free zones based on treated coke-type domestic fuels rather than raw coal. This was very important timing since the winters were proving very cold, with 20-foot snowdrifts in some parts of the country.

By now, Joyce was attending regular check-ups at the ante-natal clinic during which it was discovered that something was not quite right internally which required her to have some ring device fitted to correct her internal geometry for the eventual safe delivery of her first child. This was somewhat daunting for them both, but they were assured that it was quite a minor upset, and was expected to be one hundred percent effective. Matt's spirits were raised when, on going to meet Joyce on a shopping trip, he found her standing outside the hardware shop laughing her head off.
"What's so funny?" he enquired.
"Just look at the base of this chamber pot and see what it says," giggled Joyce.
And there, plain as plain, were the words: 'Made in Poland'. Where else would you expect a 'po' to be made? Little comic moments like this did help the waiting period along.

The happy event was delayed by a couple of weeks beyond the expected date, leaving Matt to tackle several working days with his ears on stalks awaiting an expected phone call. Eventually, he was advised that he now had a son, and that mother and baby were doing well. His colleagues nobly took over his lecturing duties to enable him to visit the nursing home. Joyce looked even more radiant that she had been during her pregnancy, when her face seemed to almost glow with hidden light, whereas the little wrinkled scrap in a shawl beside her was evidently possessed of leather lungs. All had gone well with the delivery, except that the baby, known as Steven, showed some signs of a double hernia which had to be watched. Matt and Joyce did not talk a lot – rather they sat holding hands and smiling at each other. It was too emotional an occasion for much conversation, but it was a magic moment of peace before the start of the work and worries that they both knew must lie ahead.

Parenthood came fairly naturally, and although there was much to learn on the way, the two of them were intelligent enough to listen, discuss and take on board. The double hernia was a continuing worry, but the surgeon recommended supportive bandaging until the child reached the age of one year, when it could be operated on. It proved an increasingly difficult time for Joyce because the baby would feel uncomfortable and start to cry, which would then make him more uncomfortable. However, with her nursing background, Joyce was able to minimise the soreness and keep the pain at bay to a great extent.

To his surprise, Matt found himself trying not to become too involved emotionally with the lad. It was almost as if he wanted to remain somewhat remote in case the surgery failed and the baby was lost. This was not intentional, only instinctive, and probably a throwback to the philosophical heart hardening one had to adopt during wartime when one's air force colleagues tended to disappear with a monotonously sad frequency. By great good fortune, the surgery was successful, and the baby was brought back home where he recovered steadily and launched into a normal childhood. Matt then felt able to let his love for his son flow unhindered.

However, Matt was incensed over the reaction of his aunts who, for once, were in agreement. Their first concern over the baby was whether or not it had been christened – if not, it appeared that there might be some hitch in his being admitted to Heaven should anything go awry! Matt was adamant that the first priority lay in it being cured by surgery, and that there must be something wrong with the aunts' upbringing if they couldn't see that clearly. Not surprisingly, Joyce agreed with him wholeheartedly on this. Although, like Matt, she had been brought up through Sunday School, her years in nursing had taught her that you don't just sit on your backside and ask the Almighty to cure people – you devoted your own time and energies to learn enough about medicine to do something about it yourself. Also, she had been distressed by the fact that, when her brother was warned by the police after falling in with the wrong crowd, the local church members did nothing to offer any help or guidance – they just didn't want to know. Matt's views on religion were beginning to crystallise, and he was prepared to separate what, to him, was good about it, and what wasn't, without being indoctrinated by anyone. In view of the fact that Aunt Agatha's leg was playing up again, and was not helped by her unwillingness sometimes to cooperate in accepting advice, Matt decided against entering any argument with her over the baby.

Following a cold winter, the news became grim again with the mounting of the Berlin airlift by the western Allies in order to beat the blockade imposed by Russia. It was a bold and expensive exercise, but it achieved its purpose.

When the gloomy old Head of Department at the college retired, he was replaced by the experienced senior lecturer who had been so helpful to Matt in his earlier days, and so the atmosphere improved markedly, and Matt decided to stay on for a while yet, until a proper opportunity for advancement presented itself. Furthermore, the staff's teaching hours were reduced following the appointment of a new assistant lecturer, so this made it possible for time to be devoted to refurbishing the Brockley home at weekends when Matt's workshop proved its worth over and over again.

Dennis and Val announced that they were about to move to Australia, where they would live close to the remaining members of his family. Dennis's original plans to go into the motor business, including racing, were scuppered by the burn injuries he had received, so instead he was joining, rather surprisingly, the history department of a new Australian university. His interest in history had been well known, but not to this extent. It was a sad moment for all of the Greggs to see the couple and young Harry leave, but it was their wish, and they were going to a country well suited to young people.

Matt, Joyce and Steve now enjoyed frequent visits to Brenda, Norman, and their daughter Dulcie, who were living near Bedford following the death of Brenda's mum. Norman proved to be a somewhat dour but wryly amusing character who, like so many seafarers, had found a longing for Mother Earth. He therefore trained as a land surveyor and rating officer, and was acquiring extensive knowledge of the district, its buildings, and some of its more eccentric inhabitants, stories of which kept them all amused during their visits.

In fact, life for Matt and Joyce was beginning to flower in a very pleasant way, when there was an unfortunate hitch. They arrived home one night after visiting a colleague and his wife to find that the house had been broken into. All drawers and cupboards had been opened and the contents and all their clothes strewn about the floor. In the absence of a telephone, Matt had to sprint along the road to a phone box to dial 999. The police arrived shortly after but, apart from finding an empty suitcase at the bottom of the garden – obviously earmarked to carry away the loot – they could find no trace of the thieves or the stolen items, either then or later.

There was a bit of a battle with the insurance company who wanted to shield behind the fact that a window may have been left unlocked, but some recompense was forthcoming in due course. It wasn't so much the monetary value of the stolen items that worried Matt and Joyce, it was the sentimental value of things like Joyce's engagement ring that she had forgotten to wear that evening, and the feeling of uncleanness that seemed to pervade the home after the villains had forced access to their most intimate possessions. They never felt quite the same about the house after that, and this would help them to leave without too many regrets when the time came to move on. Matt couldn't help but think, if only thieves realised the heartache they caused to decent people when they committed such crimes. But would they care if they did so? Who knows? How selfish could one get? And what were the motives driving people to do such horrible things? Were they suffering too? All these thoughts and more went round and round in Matt's mind.

Fortunately, both Matt and Joyce were resilient enough to bounce back rather than suffer from a lasting depression. Joyce, of course, was fully engaged in caring for a husband, a small son, and running a household. Although wartime restrictions on many foods were a thing of the past, shortages of some items persisted for a while, and even ration books took their time to be cancelled. There was general appreciation when bananas appeared in the shops once more, and the contents of sausages appeared less of a mystery.

Matt found an opportunity to tackle a couple of studies that had been in the back of his mind for some years. The first covered the High Boost-Low

Revs project that had started him off on the road to teaching. He had gained some more knowledge from a contact who had worked on the original project in the scientific civil service, and was therefore able to complete the whole story of power output, stemming from the early gas laws of Boyle and Charles through to supercharged piston engine performance at altitude. He was well aware that all this material was outdated following the emergence of the gas turbine, but it gave him the satisfaction of a resolution fulfilled, and he sent off a copy of his notes to the Science Museum to add to their historical records.

The second project was much more relevant to his current, and possible future, duties, because it occurred to Matt that his master plan to become a fuel technologist had had to be relegated to the back burner over the last few years owing to force of circumstances. He located his petroleum booklet that he had roughed out in Germany, blew the dust off it and set about bringing it up to date. It meant laborious typing, and draughtsmanship regarding the illustrations – there being no computers or word processors in those days – but he pressed on to eventual completion. Then came approaches to potential publishers. After a couple of rejections, one publisher was found who offered useful advice, and agreed to print the work as part of its university series. This took several months to accomplish, but Matt was thrilled to bits when he held a copy of his very first book in his hands. It was, he thought, the nearest thing to giving birth.

A congratulatory party was held at Brickhill, and the family rejoiced in Matt's success. While there, Matt took the opportunity to clean up Mike's sports car, and put it on the market. Matt would have loved to have kept it, if only in memory of Mike, but a sports car that suited a dashing young fancy-free pilot did not quite square up with the needs of a married couple and their infant son. The car went to an enthusiast who, Matt was convinced, would give the car the tender loving care that Mike would have approved. The purchase money was put aside by Dad to share between Val and Matt when they were ready to buy their own first cars.

The sense behind this decision became evident when Joyce announced to Matt that she was pregnant again. All prospects of writing in Matt's 'spare' time were now dislodged by the need to prepare accommodation for the new arrival. Young Steven was now two years old, and wanted to know when they were going to the shop to buy the new baby. He was told that it took a little while for the order to go through, and was encouraged to join in the movement of his cot – and even some of his toys – into the spare room set aside for a new nursery. This time, the child was carried without difficulty, and a baby girl, Avril, arrived as a home birth.

Despite the additional parental duties, Matt attended to another topic that had exercised his mind for some years. This concerned the inconvenience of having to adjust both hot and cold water taps over the wash basin to achieve the temperature of the water you wanted, and then having to readjust both of them when you wanted to change the speed of filling the basin. He therefore sat down and designed a composite mixer tap that would permit temperature and flow rate to be controlled separately. He applied for a patent for his idea but was dismayed to find how expensive this would be if he proceeded. Furthermore, he discovered a flaw in his design, in that it would be possible for the hot water derived from an open storage tank in the loft – subject to contamination by dust (or worse, when he found a dead mouse in the tank!) – to pollute the incoming mains water, so he abandoned the idea. In any case, it had proved a useful exercise in clarity of writing – essential apparently because he was told that some companies employed people to examine patents solely with a view to spotting a weak point where they could step in and exploit the idea for themselves. This thought disgusted Matt, and he was glad that he was in academia and not that type of industry.

It was no real surprise when the news came through that Aunt Agatha had died. She had recovered successfully from a leg amputation, but succumbed later to a heart attack. What was surprising was that Belinda expired a few months later since the family thought that she might blossom with no one around to criticise her continually, but it seemed that the arguments with her sister had given her the spice she needed for life, and she was lost without it.

It came as an enormous relief to every motorist in the UK when petrol rationing ended, and Matt decided that he would invest in a car as soon as he could scrape up the necessary cash to add to that available from Dad.

Chapter 12 – Oil (1951 – 54)

By now, Matt and Joyce had settled down to a gentle and enjoyable married life with their two children, Steven and Avril, developing through the normal childhood activities involved with birthdays, Christmases, pre-school attendance and holidays as well as the downside of chicken pox, dental treatment and the odd disagreements with neighbouring youngsters. There is always a thrill in witnessing children learning about themselves, their family and their surroundings, and Matt and Joyce found this deeply satisfying – although traumatic at times!

As part of the children's education for life, they had taken on ration strength a four-footed friend in the shape of a young all-black cat who answered – when he felt like it – to the name of Toby. They considered that cats were fascinating creatures in that they clearly needed human support for a comfortable non-feral existence, although they invariably behaved as if they didn't. Toby had a bit of a problem adjusting to his new diet, but once the appropriate 'moggydins' had been discovered, Matt encouraged its provision with the words "That'll put hair on his chest!" And so it did.

The cat was also bemused by the sheer strangeness of human behaviour, no doubt thinking, 'They walk about on their hind legs all day, then take their fur off at night! Weird. No self-respecting feline would behave that way.' But as long as his dinner pail was kept topped up, Toby graciously accepted their eccentricities. After a few accidents requiring the application of carpet cleaner, the cat reluctantly decided to become house-trained. What proved more difficult was to curb his excessive delight in bringing indoors dead mice, birds and other trophies for the family to enthuse over – especially when they found them in their shoes! Came the time when a visit to the vet was called for in order, as Matt explained to the children, for the cat to have his tonsils out.

Further good news arrived from Brenda and Norman regarding the birth of their son, Geoffrey. It appeared that Norman had been suffering from stress for a while, but was now in much better shape.

It was at this point that Matt felt able to join the motoring fraternity. A colleague had offered him his old *Morris Ten*, partly because the rear seat and tail section were wobbly, and he had no idea how to fix it. Matt brought it home in triumph, and they all stood around gazing at it in wonder. Then Toby turned up and, having sniffed the tyres to his satisfaction, promptly settled himself down on the warm bonnet. The family took this to mean that he had given his approval for them to keep it. The cost of the car was easily covered by Matt's share of the funds from Dad. Matt was able to secure the rear seat firmly, and found the tail problem to be due to the fact that the body structure comprised a number of wooden struts, rather like half-timbering but internal, and that some of the struts were loose. By making suitably shaped metal brackets and screwing them onto the wood, stability of the rear end was regained. Matt made other improvements by replacing the headlamps and running boards, and by fitting a heating pipe under the dashboard, tapping in to the hot engine coolant.

One of their first forays, of course, was to Brickhill, but this was before the motorway era, so a journey of this length and complexity represented quite an adventure, entailing at least one stop on the way for coffee and sandwiches. When an AA patrol pulled in alongside, Matt took the opportunity to join. He felt that the combination of an old car and a young family warranted some extra support. He remembered one of his friends who had an old banger, with a floor so corroded that it had to be reinforced with corrugated iron sheets, which looked so decrepit that passing AA and RAC patrols used to turn and follow him, waiting for the inevitable accident to happen.

One hill on the Brickhill route was so steep that the car could not handle it when fully laden unless it had gained sufficient speed at the start, so Matt had to hang back until the approach road was clear before taking a run at it. With such a relatively simple gearbox, double declutching was the order of the day, but Matt had mastered this with the RAF's 3-tonners. In those days, of course, it was necessary about every few thousand miles to give the engine a top overhaul by taking the head off and decoking it, also grinding in the valves and attending to other bits and pieces. It all meant work, but enabled to motorist to get to know his engine better.

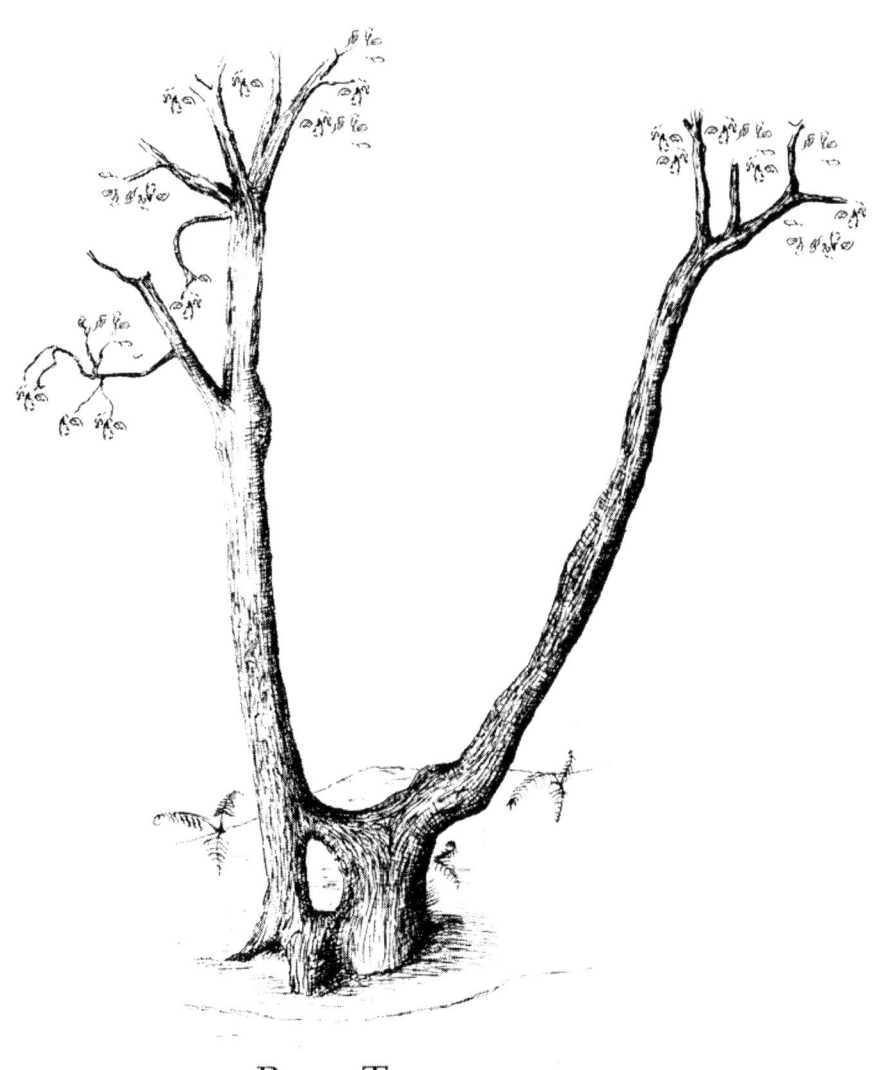

Re-enTree

One of their favourite pastimes was to visit nearby Keston with its ponds and Caesar's Well, and Daddy always made a point of taking them to see the Spitfire and Hurricane aircraft mounted at the gates of Biggin Hill airfield where Uncle Mike had spent several active months. He also pointed out Downe House, home of the great Charles Darwin. Once he took them down the hill to Westerham to show them the location of Chartwell, home of Sir Winston Churchill, who had piloted us through the Second World War.

When the family visited Brickhill, their delight was to wander through the woods – the favourite haunt of Rex – and to have a picnic at the foot of an unusual oak tree in Aspley Heath which had strangely managed to arrange for one of its branches to curve inwards and grow back into the trunk. Matt named this as the Re-enTree, and it became very special to them all. Matt's only fear was that it would be just their luck for the confounded thing to be felled during some over-enthusiastic replanting scheme, or perhaps blown down in a high wind, but it grew doggedly on regardless (and is there to this day!). There was also a tree nearby with two similar trunks arranged in a very definite 'V' pattern that Matt called the VicTree, and another which was biased to one side which he dubbed the A-symmeTree, but the Re-enTree was their favourite.

While the grandparents cared for the children, Matt and Joyce joined the crowds at the Festival of Britain exhibition on the South Bank, to marvel at such wonders as the Dome of Discovery, the Guinness Clock, and the Tree Walk. They were particularly impressed with the Skylon, a massive cylinder, tapered at each end and suspended vertically by a system of wires. Matt was intrigued to note the use of strain gauges on the wires to check the loading – the first time he had seen them used outside the aircraft industry. Our happy pair rounded off the day with a romantic dinner at a good quality restaurant. It cost a packet but, as Matt said to himself, it was a precious moment, so hang the expense – give the cat another goldfish!

While they had some time to relax, they arranged to take a holiday. They chose Folkestone, where Matt had memories of the magical fairy-lit zig-zag path leading down from the bandstand to the beach, and equipped with numerous romantic arbours ideally suited for loving couples. Unfortunately, of course, he had been too young to take advantage of these facilities before the war, and it was a bit late now that he was a married man with a family, but better late than never.

On their return, Joyce was deeply saddened by the news that her mother had died following a stroke, but this time Matt was there to comfort her, and to help make all the arrangements, including the mountain of paperwork that seems to accompany such events. His skill as a trained touch-typist once again proved most useful. Joyce had been very close to her mother, a kindly soul prematurely aged because of an erring husband whose weaknesses led him to drinking and gambling away the few funds they could muster. The root cause probably lay in the trauma he had undergone in the trenches during the First World War, but frankly his passing had resulted in both sorrow and relief. Joyce's mother had soldiered on to bring up the children as best she could and, because of her nature, she had become the receptacle for the troubles of everyone around her.

There was further dismay when Matt arrived home to find that, during an uncharacteristic bout of energy, Toby had just wandered into the road and been struck by a car. Joyce was making him comfortable in a padded box, and they then made a beeline for the vet. There was relief all round when no bones were found to be broken, only dislocated, and after being sorted out, the cat was taken home to be nursed. The next problem was that he could neither eat nor drink, so chances of recovery appeared slim, but Joyce then had a brainwave. She dipped a spatula in water and dabbed it on Toby's nose. As a natural reaction, the cat applied his tongue to his nose, and thus took on board a few drops of water. After two such applications, he seemed to cotton on to this as his form of drinking for the moment, and Joyce spent the next hour hydrating him in this way. This seemed to be the hinge on which his fate pivoted between returning to health or joining his ancestors. As Matt put it: "Toby or not Toby, that was the question." But he still had eight lives left.

The arrival of some modest, but useful, royalties from his book galvanised Matt into contemplating a serious move to study the subject in greater depth, so he made application to a number of firms in the oil industry, enclosing a copy of his book. One company, it transpired, was in the market for new staff, so they offered him an interview at their London offices. Matt dressed very smartly that day, and Joyce surveyed him critically to ensure that he appeared at his best. She felt proud of him, and kissed him to wish him good luck.

In the train, Matt ran over in his mind the likely questions that he would be asked, and the answers he could proffer. Just as importantly, he mentally listed the questions that he wished to be answered by them – he was sure that this approach would count in his favour. His confidence was strengthened by the news that Winston Churchill was once more Prime Minister. He was certain that Dad would be very pleased.

The interviewing committee members were courteous and welcoming, although it soon became apparent that they were principally looking for people to go out to their Middle Eastern refineries. Matt made it quite clear that, having already served in Europe for some time, and now having a young family to support, he wished to remain in this country. The interview might have ended there, but the interest and diligence that Matt had demonstrated in preparing his book impressed the committee sufficiently for them to offer him a position as Assistant Engineer in their research station just outside London. Matt jumped at the chance, not just because the salary was significantly above his present level, but because it would enable him to get to grips with high quality research in his chosen specialisation.

Joyce was delighted for her husband, although she pointed out that he would have to rise particularly early every morning to catch the train to Waterloo, and then another to the research station, since it would be too far to take the car day after day. They agreed that it was worth the effort and that, if it did prove difficult, they would have to consider moving to the other side of London. In this event, the oil company had promised to offer the services of their solicitors.

And so Matt extricated himself from the college, promising to maintain contact with all those who had helped him along. He duly presented himself at the research station, where he was advised that he would initially be involved in a research project concerning the distribution of air-fuel mixtures to the individual cylinders of a large automobile engine. The engine was installed in a sound-proofed test cell, and was instrumented up to measure the mixture strengths arriving at the cylinders so that the distribution pattern could be measured, together with the effects of operating conditions like throttle opening, engine speed etc. He was to work with Pat, an Assistant Engineer of much the same age as himself, and both of them were to be supported by Fred, a cheerful fitter of considerable skill in his trade and highly developed sense of humour.

The work proved extremely interesting, and Matt soon became aware of the weaknesses of engine systems relying on carburettors feeding an inlet manifold. It emerged that both air and the vaporised fuel components would distribute evenly and be no real problem, but some of the fuel remained as liquid during its passage through the manifold and, being much more dense than the air and vapour, it could get thrown about within the manifold depending very much on the geometry. This meant that whereas some cylinders were receiving correct air-fuel mass ratios of about 15/1, others were rich at 13/1 and some weak at 17/1. What surprised Matt was that the weak cylinders ran hot – he had imagined it to be the other way round – but Pat explained that because the flame speed was lower, burning was still occurring on the exhaust stroke, the unused heat going into the cylinder and then back into the following charge. The rich cylinders on the other hand, tended to smoke. The whole effect was a sorry sight of uneven combustion, loss of fuel economy and possible early knock that would limit power output. On reflection, Matt remembered that even the splendid Merlin experienced slight excesses of fuel, and its TEL additive, finding their way into the front starboard cylinder.

It was this experience that started Matt's aversion to carburettors. Some of them were very ingenious in their design to introduce the fuel into the air stream, but they suffered from a number of problems including a natural, unwanted enrichment at altitude, and the fact that, being controlled by a float chamber, they could not operate well when inclined or in upside down flight – although this snag had largely been overcome by the insertion of an additional orifice in the floor of the float chamber. The main point though, was that however clever the design of the carburettor, the problem of maldistribution arose once the mixture had left on its way through the inlet manifold. This problem persisted even when the improved system of continuous injection into the eye of the supercharger was adopted, although the useful cooling effect of 25°C due to fuel evaporation was retained. Clearly, the best solution lay in fuel injection to individual cylinders, just as the Germans had used because of their background in diesel engines, and to use a charge cooler either after the supercharger or between its two stages. The importance of cylinder injection was proved beyond measure when the whole question of atmospheric pollution from exhaust products became such a vital issue in later years.

Matt became so fascinated with the subject that he took his notes with him to read in the train on the way home. After all, why waste valuable time surveying people's back bricks and whitewashed privies? Eventually, after writing the necessary company reports on the work, Matt felt that he could take it a bit further, with some more theory thrown in, to prepare a thesis that might warrant a higher degree. With Joyce's support, he organised his home duties to be as efficient as possible, and then settled down to work. He discussed this with his employers and was a bit put out to find little enthusiasm, the policy being that, although they would permit him to publish their results in this way, since he was already working satisfactorily there, an additional paper qualification was not really necessary. But Matt had other ideas. Always in his mind was the possibility of an academic life, and so he persevered, and ultimately submitted the appropriate number of copies of his thesis for a Master of Science degree to the University of London as an external student.

While awaiting the hoped-for interview, he then concentrated on seeking part-time lecturing in Thermodynamics at local colleges, and was delighted to receive offers that would have occupied all five weekday evenings plus Saturday mornings! He had to rationalise these, of course, although he had the advantage of preparing one lecture and giving it several times a week, instead of having to prepare for each one separately.

* * * * *

In 1952 the country was plunged into mourning with the death of King George VI, and people began to realise more clearly the difficulty he had had in accepting kingship so suddenly after the abdication of Edward VIII.

Gloom was not dispelled by news of the disintegration of an aircraft at the Farnborough Air Show, with the loss of 26 lives.

It was during this period that it dawned on Matt that there was much more to technical writing than he had first supposed. Certainly he had done his homework in collating and checking the information, and expressing it as clearly as he knew how, but there was the question of writing style that he had not studied so far. When he did so, he found that although some of his instincts had been correct, he had much to learn about the structure of sentences. For example, "For both types of piston engine…" would be better expressed as, "For piston engines of both types…" And, "For reasons of safety, low flammability fuels are used," should be revised to, "Low flammability fuels are used for reasons of safety."

Also recommended was the need to keep the writing concise. This was brought home to him later, when he compared one of his opening sentences: "The emergence of mankind" (7 syllables), with Dr Bronowski's television presentation, "The ascent of man" (5 syllables). Those two syllables made a world of difference between a forgettable comment and one with ringing tones that became famous.

Matt could not clearly define precise rules for good writing style, but by wide reading of the appropriate books and articles he found he was absorbing an improved technique almost subconsciously by a process of osmosis. At all events, he hoped that his first fuels book would soon be lost and forgotten, and that he would have an opportunity later to prepare a much improved edition.

The day of the university interview arrived, and once again Joyce supervised his best-suited appearance, and also made a point of polishing his shoes herself in order to take a maximum part in his support. Matt was able to drive into London fairly easily – severe traffic congestion was still a problem of the future – and to present himself at the appropriate college venue. He was interviewed by a Professor and a Senior Lecturer, who were both thoroughly conversant with the subject, and posed questions that were searching but fair. Their main concern, it appeared, was that the bulk of the work presented had been undertaken by the candidate, who must also show a sufficient level of intellectual merit. Matt was then asked to step into an adjoining office to await the verdict. After a period of about two years that must have lasted at least ten minutes, he was ushered back in to meet the outstretched hand of the Professor, and be congratulated on having successfully achieved the award of an MSc (Eng) – although he would not be permitted to use these letters after his name until receipt of official confirmation.

Matt, of course, was overjoyed, and searched high and low for a public telephone so that he could give Joyce the good news without delay –the luxury of a telephone by now having been installed at home. He then made a beeline for the refectory because he felt the need for a cup of coffee before facing the return drive home. Once there, they were faced with a question from the children as to why Mummy and Daddy were dancing round the kitchen like that? Toby, of course, gave up in disgust at such antics, and sought a siesta on his favourite position on the car bonnet. The news was received with delight at Brickhill, and the suggestion that they all meet up there as soon as possible for a celebratory visit to some local hostelry. Dad advised of a particularly good venue at Newport Pagnell where you could partake of a splendid meat dish, of which he couldn't quite remember the name but it was something like Tournedos Medici. They would love it, he assured them. And they did.

Events in most families seem to follow some sort of sine wave, with the peaks of happiness interspersed with troughs of sadness, but this moment was undoubtedly a peak of some magnitude.

On his return to work, Matt was greeted with congratulations from both Pat and Fred, and it was arranged for all three to meet at a nearby pub that evening where a game of darts was interspersed with tankards of a local brew. It was no real surprise that Fred, with his practised manual skills, metaphorically beat the pants off them both, but a good time was had by all.

Having completed the mixture maldistribution project, Matt was then directed to survey methods of measuring the sizes of droplets from a nozzle spray. After that, Pat and he were started on a programme to explore the rate of build up of carbon on a diesel nozzle. This was a bit unsatisfactory because, however carefully one disconnected the nozzle and removed it from the cylinder, the trumpet of carbon could break off. Also, Matt began to feel that, despite the genuine interest he was finding in these researches, he would appreciate a chance to choose his own project rather than always be directed from on high.

"He certainly puts it through its paces!"

Fred, meanwhile, gave continuing support with any work involving hand tools, and Matt was very pleased to learn from him several wrinkles regarding fitting and brazing. Fred also had the job of bringing from Stores the cans of special fuels required for the tests, and consequently described himself as the 'Jungle Juice Wallah'.

Matt made quite sure that he would not overlook their tenth wedding anniversary, and he managed to organise a memorable outing with Joyce, together with a gift of a delicately designed wristwatch.

The most dramatic event that took place in the research laboratory during this period occurred on the day that a single-cylinder diesel test-engine was delivered. Once the wrappings had been removed, Pat gave the crankshaft a swing just to check that it was free, with no nuts and bolts left in the cylinder. To everyone's amazement, the engine fired and started to jump round the lab! After discussion, they realised that it was the protective oil in the cylinder that had ignited spontaneously in the air heated by the compression. This proved to be another watershed for Matt because it made him realise the difference between ignitability and flammability, a feature that was to become a cornerstone in his subsequent lecture career. (See Appendix 3.)

Country wide there were more disasters, with the loss of life in the East Coast hurricane and the sinking of a car ferry near Ireland. Then Queen Mary passed away. But spirits rose again with the crowning of Queen Elizabeth II, in June 1953 and the conquest of Everest by Hilary and Tenzing at about the same time. Some time later, it was decided to ground all the Comet jet aircraft in view of the spate of crashes.

The year ended on a happy note for the Greggs as Dennis had been granted sabbatical leave, and he arrived, with Val and Harry, to conduct a survey of the teaching of history in British universities. Everyone was delighted to welcome them – particularly Mum and Dad. To crown it all, rationing (of meat, cheese and butter) came to an end at long last.

Chapter 13 – Canals (1954)

Matt was absorbed with the research projects of the oil industry, and he found that the pay and conditions were good. With Joyce, he had built up a pleasant social life that complemented their own family environment. Steve was established in a nearby school, making reasonable, though not outstanding, progress, and Avril was just starting. So all in all, life seemed pretty comfortable, particularly when compared with their memories of wartime deprivations and dangers. However, there were these constant naggings in Matt's mind about the urge to pilot his own career by selecting research projects of his choice, and by finding opportunities to disseminate what he had learnt. There was just no time to devote to part-time teaching at local colleges, much as he would have liked, although he could find some satisfaction in that direction by technical writing, especially as he could do some of the preparation during the train journeys.

It was in this frame of mind that he noticed and homed in on an advertisement in the technical press for experienced academics/researchers to support the development of a newly-formed university in the Beds/Bucks area. An appointment such as this would not only give him the research independence and teaching aspects that he sought, but was also within striking distance of Brickhill. Matt and Joyce talked this over at length, and decided that it could be a very wise move indeed since not only would Matt be more fulfilled in his job, but they could probably arrange to rent the cottage at Brickhill, and so free-up the London property for Dad to sell. Furthermore, although they found London a multifaceted city, they would prefer to live closer to the countryside, and bring up their children there too. Matt had a quick word with his father, who shared their enthusiasm, about these possibilities, so he went ahead and submitted an application.

An interview was duly arranged at a London office, and Matt played carefully his cards regarding lecturing experience, management in the services, and technical writing in the shape of a few articles and his fuels book, hoping against hope that his somewhat amateurish writing ability would not be noticed. In fact, the committee did glance through the book, but their attention centred on the figures rather than the text and, since Matt had the instincts of a draughtsman and the ability to illustrate his work evocatively, they were clearly impressed. Matt enjoyed the interview, and felt confident that he had given it his best, but it emerged that there would have to be a fairly extensive delay for a decision while other candidates were interviewed. Matt explained that this presented no problem since he was planning on taking a holiday very shortly.

In the meantime, Matt shared the spare hours he had between minor repairs to the house – in case it would be going on the market – and revising his fuels book to improve the style. Fortunately, the publishers agreed to a second edition, and Matt was able eventually to hand over the finished product, reducing his commitments by one.

Throughout this time, Matt and Joyce had been careful to ensure breaks for the children and themselves to have some time off at weekends to relax. Joyce had become engrossed in sketching and water colours, having already produced items worthy of framing and hanging. Steve had acquired an old bike, but was happiest when exploring the fauna and flora of some nearby woodlands that had so far escaped the developers. One notable occasion arose when Steve and his mates embarked on a sortie to collect conkers and string them up for a tournament – this was ages before a nanny state decreed that safety goggles should be worn! A phone call was then received from the mother of Vernie, Steve's best friend, to the effect that all the lads had had their tea there, and that Steve was about to leave for home. And by the way, he had been the winner of the tournament! As he returned home in triumph up the garden path, Matt could not suppress the comment, "Here our conkering hero comes!" Avril, on the other hand, was beginning to take more notice of her own appearance, and to feature in amateur dramatics at school – and at home when the occasion demanded!

Eventually, the day arrived with a letter offering Matt a Senior Lecturer post at the university, with the special responsibility for lecturing and researching in Fuels and Lubricants. Although the salary was barely higher than his present level, they felt that all the other considerations made the offer acceptable, so Matt set the wheels in motion to resign from the oil company, take up his new post as soon as practicable, and prepare the Brockley house for sale on Dad's behalf. The oil company did not want to lose him, and tried hard to dissuade him from leaving, but Matt explained his feelings and aspirations, so he was able to leave under amicable conditions. Mum and Dad were delighted that their grandchildren would be so near, yet in their own home.

The priorities that followed were concerned broadly with tapering out from London, and tapering in to Brickhill. The Brockley home was put on the market, and they were advised to keep the place as clutter-free as possible so that potential buyers could imagine the rooms housing their own artefacts rather than those of the present occupiers. They did not have to be told to hide anything valuable in the event that someone might 'case the joint' for a subsequent burglary. The tenant in the Brickhill cottage was given the appropriate notice, and Matt started to hand over his responsibilities at work to Pat.

A kindly neighbour provided a child-minding service so that Joyce could join Matt on his farewell party, then it was time for the removals men to arrive with their tea chests for packing and loading. Toby was not best pleased to be caged in a cat basket, but eventually settled down to sleep. The family drove behind the van initially, but led the way over the last few miles. They had to move in with Mum and Dad temporarily since the cottage notice period was not yet up. However, the Brockley property was disposed of satisfactorily. All in all, a reasonably successful operation, even in the opinion of Toby who had visited Brickhill before, and rapidly reclaimed his favourite spots in the house and garden.

Matt had arranged for a few days' break before taking up his new post, so Joyce and he sorted plans for a holiday. Matt realised that Joyce and the children needed a change just as much as he did, and they eventually settled for a week of narrow-boating on the Grand Union Canal. Messing about on boats had always appealed to Matt, and Joyce looked forward to the sheer peace of watching the countryside slide by at a gentle pace, and they would keep the cooking chores down as much as possible by eating out. The children, of course, were thoroughly excited at the prospect of living on a boat.

An early decision had to be taken regarding Toby since he would give not a thank you for being incarcerated in a steel box for a week, with all those neat hunting grounds drifting by so close and yet tantalisingly beyond reach of paw and claw. Besides, he would probably only keep the family awake at night by stamping about the deck and stropping his backbone against the mast. Mum and Dad were just reassuring them that they would look after Toby, when the telephone rang with a message from Val to say that they would shortly be returning to Australia, but they had some time before leaving so could they all arrange to spend some days together? The upshot of this conversation was that Dennis, Val and Harry would love to join them on their canal holiday, and would Matt mind making all the arrangements while they sorted out their packing?

Joyce was delighted at the thought of being in the company of another woman, and sharing the cooking duties with her. Avril was also keen to be involved in cookery, and insisted on making cakes for her dolls, while Steve cherished the thought of poking about in the water searching for wildlife. Matt much looked forward to man-to-man talking with Dennis, and learning from him some of the art of navigation, with which Dennis had been concerned to some extent as a member of a bomber aircrew. The children also wanted to hear what schooldays were like in Australia, and had Harry seen any real live kangaroos in the wild?

A week's cruising was booked on an eight-berth boat at a marina near Leicester. Fortunately, the activities of moving to Brickhill had occupied part of a waiting time that would have otherwise been frustrating for the children, so their preparation for departure had to be undertaken over a very few days. However, the copy of the inventory sent them was remarkably comprehensive, including bedding, kitchen utensils of every description, cutlery, crockery, cleaning equipment, coat hangers and doormats, together with navigating items comprising boarding plank, boat hook, mooring spikes, hammer, windlass and hose. Also, a small TV set was available for a modest fee. As Matt remarked, "For all who come to pass upon still waters – they shall not want." Exactly so. The excitement blossomed further with the arrival of Dennis, Val and Harry, but Matt and Joyce deliberately kept as low a profile as possible so that Mum and Dad could have most contact with their daughter and her family.

Very little sleep was achieved that night, but everyone was up betimes, and the two cars then departed north north west along the motorway, and parked at the marina. There they were thrilled to find their 56 ft cruiser already dusted, fuelled and watered up ready to go, and were impressed with the interior layout based on all the doors fitting snugly in two places in order to close off wardrobes, etc by day, and give individual-cabin privacy by night. She had two separate batteries, for engine and general services respectively, together with four bottles of petroleum gases for cooking and central heating. A BMC 4-cylinder diesel engine provided propulsion, and the tank contained 15 gallons, plus one gallon for luck, of diesel fuel as the expected average consumption for one week, although there was extra fuel in the tank if required, at cost. On the waterways, in contrast to the road, diesel fuel is less expensive since it is tax free and consequently dyed (pink) for identification.

They learnt that, on the canals, distance cruised is usually measured in terms of lock-miles, given by the formula:

$$\text{Distance in lock-miles} = \frac{\text{Locks negotiated}}{2} + \text{Miles travelled}$$

Since a comfortable average speed for a cruiser is three lock-miles per hour, this represents ten minutes per lock. Hence this speed over one hour could extend from one limit of a flight of six locks negotiated with virtually no distance covered, to the other limit of a three-mile stretch of lock-free canal, with any combination in between, although the maximum speed is limited to four miles per hour, otherwise the wash damages the banks. One of the potential hazards for anyone of Dennis's height was the clobbering of one's head on the hatch covers when negotiating the steps to and from the deck, but Dennis had foreseen this and brought with him his old flying helmet. Climbing in and out of aircraft over the years had left its impressions in his mind, and sometimes in his cranium!

The marina staff gave as much instruction as they could prior to departure, including the four cardinal rules that the crew must NEVER:

"No. I think you're wrong! It's a Swa......"

1. Tie up the boat when going 'down' a lock – an 18 ton cruiser suspended solely on a couple of ropes is not a pretty sight.
2. Leave a lock without closing all gates and paddles – a dewatered canal is no use to man or boats.
3. Leave a windlass handle engaged on a paddle spindle, since this could be guaranteed to fly off the spindle as it slips the ratchet and probably clobber at least two crew members, before disappearing for ever into the depths below.
4. Stand up on deck looking aft when passing under a bridge. Masonry and metal are far harder than the human skull!

But there was still much to learn on the way. Dennis cast off, Matt gingerly aimed the boat to the western outlet of the marina, and they found themselves well and truly launched. After a few careful miles, it was deemed time to tie up for coffee, and here they encountered their first lesson when the mud was churned up and the propeller ceased to rotate. Matt and Dennis discussed the situation and opened the weed hatch to plunge their hands into the cold cocoa-like water to drag up the greenery so lovingly entwined around the prop. They then came to realise that, as the cross section of the canal is roughly semi-circular, the water is very shallow at the banks, whereas the prop needs the deep water of mid canal. In future, therefore, they would:

1. Nose into the bank and get one of the crew to step ashore with the hammer and both fore and aft ropes.
2. Hammer in one of the mooring spikes and tie up the nose rope to it.
3. With the throttle in neutral to keep the prop still, pull in the stern to the bank and tie up that rope to another mooring spike, ensuring hammer then returned to boat.
4. Switch off engine.

They enjoyed their coffee after that, feeling that they had cracked problem No. 1. But, as Joyce pointed out, what about problem No. 2, i.e. leaving the mooring? "Oh yes, there's always that," replied Matt, so the two men went back to the drawing board and came up with the following drill:

1. Ensure throttle in neutral and switch on engine.
2. Untie aft mooring rope and return it to boat, not forgetting the spike.
3. Push the stern out to mid canal using the mooring pole.
4. Untie forward mooring rope and climb aboard with it and spike.
5. Select reverse and bring boat in line with mid canal.
6. Select forward and proceed.

Just when the two men were beginning to feel a little smug at their mastering of waterways navigation, they were faced with another situation requiring further thought. On approaching one of the many bridges on that part of the canal, they met an oncoming boat to which they had to give right of way. They already knew about passing on the right, but Matt found it difficult to control direction at the slow speed that was essential for such a manoeuvre. Again, with their two brains wrapped around the problem they realised that, despite the slow overall speed required, they should give the throttle a burst so that the accelerated water impinging on the rudder gives it greater effect. After some practice, both men became reasonably skilled at this, and were able to instruct the ladies, and the children, when they wanted their turn at helming.

There were other wrinkles they learnt, too, like dropping the anchor at night when moored anywhere near a pub, since late-night revellers seemed to think it a good idea to untie mooring ropes and let sleeping boats drift loose. Also, while waiting to enter the lower level of a lock, Matt imagined that the water would push the boat backwards when it was released, so he should be ready to select forward to cope with it. In fact, the water emptied at the bottom of the canal section, circulated upwards and slammed the boat into the lock gates. No wonder these boats were all fitted with massive fenders at their bows!

Despite the fact that this was a holiday, Matt could not resist the temptation to keep a log of their cruise, with a view to writing it up later, complete with photographs and map, to remind them of their adventure in later years. This was in no sense a chore, since he was now used to writing, and found non-technical scribbling quite pleasant, even therapeutic. The log covered their departure from Kilworth through Yelvertoft, Napton, Long Itchington, Royal Leamington Spa to Warwick and return. This included the 1,528 yards of exciting spooky darkness with all lights on as they negotiated the tunnel at Crick. Apparently, problems of quicksands delayed the opening of this tunnel until 1814. Water dripping from the tunnel roof was countered by the fact that Val had thought to bring along an umbrella.

The men were intrigued to note that headlights of approaching boats appeared to be red in the far distance, viewed through the exhaust gases dispersing only slowly from the tunnel, but gradually turned to white as the distance shortened. "Of course," they realised, "it's because red is more penetrating, like you see the colour of the low sunshine through the atmosphere in the evening." There was a similar experience in the 2,042-yard long Braunston Tunnel, which was much drier. This was a bit wiggly, probably also because of the quicksands problem, although it was opened earlier in 1796. The peace and serenity of gentle canal boating was brought home to them when they cruised under the main highway and witnessed frenetic motorists hurtling along at well over three lock-miles per hour. Mad fools!

Stimulated by his notes for their cruise log, Matt explored the literature of the history of the canals themselves, and his final description appeared as follows:

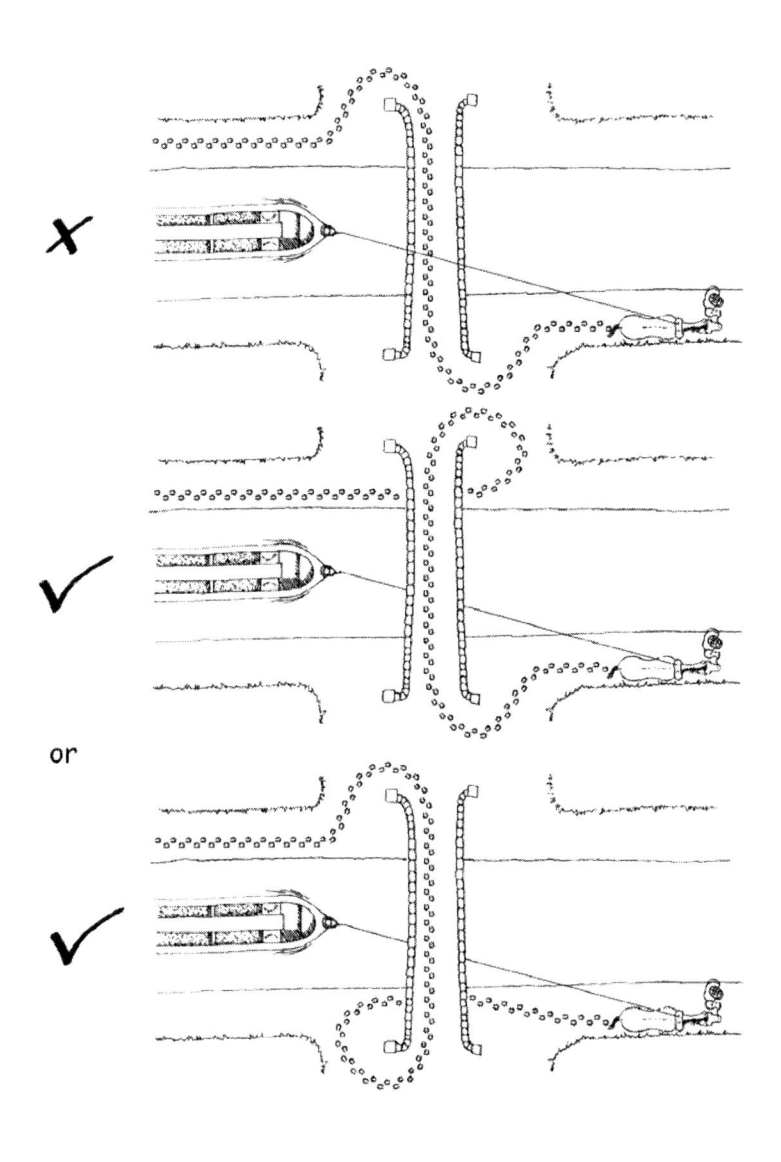

The UK Canal System

The UK canals mark a link in the chain of transport development, since they became necessary, and proved commercially successful, due to the inadequacy of the highroads, and then in turn succumbed to the spread of the railways. The building of British canals is generally considered to have started way back in 1745, and to have reached a peak in about 1791-4, terminating in 1835. One can imagine the mixed bag of itinerant navigators ('navvies') attracted by the comparatively high wages, digging their way through the countryside, and often living off the land – no doubt to the dismay of some local farmers. The digging was all done by hand, and then puddled clay was used to seal the canal bed.

Water was supplied from rivers or reservoirs on the summits, and let through the locks at each end, the canals being made narrow to conserve supplies. The narrow boats that evolved were originally unpainted, and crewed only by men but, as wages declined due to railway competition, crews could no longer maintain their families ashore, and so mum and the kids had to come along too. The respective mums lost no time in making their boats look a bit more like home, and so created the colourful decorative art form that has survived ever since.

Boat propulsion was originally by man towage until the horse took over in 1800, for which the men were no doubt duly grateful. Many a man has been known to follow the horses, but here the tables were turned. The typical narrow boat was designed for a payload of about 30 tons, but there were even express packet boats for passengers, commanding the best horses and right of way over all other traffic. An interesting problem arose when the tow horse needed to cross from one towpath to that on the opposite side of the canal, because if the horse crossed directly, the tendency would be for the boat to try to follow the rope over the bridge! Ingenious but simple solutions included the split bridge, incorporating a gap between twin cantilevered arms to allow the rope to pass through. Alternatively, the turnover bridge was used, where the horse either first passed under the bridge before crossing, or else crossed and then proceeded from under the bridge, the projecting brickwork in each case being protected against wear from the tow rope.

The lack of towing paths in most tunnels made necessary the slow, dangerous, exhausting and rather undignified system of 'legging', i.e. with two men lying on their backs either across the foredeck or on two projecting 'wing' boards and propelling the boat by moving their feet along the tunnel walls, with the hazard of falling into the water and being crushed between the boat and the wall. This technique persisted at Husbands Bosworth and Saddington even up to 1939. Various schemes of endless ropes and tugs (steam, electric and then diesel) were tried until the self-propelled narrow boat (first steam, then diesel) eventually emerged, towing a dumb 'butty' boat to double the payload without doubling the costs. Commercial carrying virtually ended by 1970, but the present-day canal pleasure cruiser has inherited the general principles of narrow-boat design, decoration and diesel.

But there was more to this than mere scribbling about one piece of history, and Matt had a revelation almost as strong as his conversion from aeronautical engineering to fuel technology, for he realised that he derived such great pleasure from studying history and then presenting it in a light-hearted manner that people might find easily readable. He felt this was the perfect relaxation for him from the formal technical writing with which he was normally involved at work. So, with the help of the AA publications, he roughed out the following:

An Outline History of the South Midlands

This area is centred within the general confluence of the counties of Leicestershire, Northamptonshire and Warwickshire. The scenery is generally pleasant with

Leicestershire boasting some of the richest grazing land in England, whereas Northamptonshire was formerly called the county of squires and spires. Warwickshire is particularly apposite to the canal story since the centre of the whole system is located at Birmingham. In fact, the centre of the whole country is claimed by the inhabitants of Meriden (Warks) to be marked by their medieval cross, a point contested by the locals of High Cross, the Lillington Oak, and also of Weedon (Northants). Fortunately, as far as we know, they haven't come to blows over it, which is somewhat surprising in view of the depressing list of battles and other downright unfriendly acts extending throughout the history of the whole region, which have collectively prompted the title 'The Cockpit of England'. Even though the years have mercifully dimmed the recital of such historical gems at school, whilst trying to concentrate on the more evocative 'Tiger Tim's Weekly' concealed beneath the desk, names like Bosworth, Edgehill, and Naseby still stir a chord.

Warwick

This county town is guarded by a castle built primarily in the 14th and 15th centuries – little wonder that it took that long without JCBs and mechanical diggers available. It has been inhabited continuously to this day, mostly by successive Earls of Warwick, ravages of the Civil War happily being avoided by the Parliamentary leanings of the then owner, a farsighted Lord Brooke. The castle's appearance from across the Avon prompted Sir Walter Scott's description as "the finest prospect in Europe". Following the Battle of Poitiers, French prisoners were held in the Caesar's Tower, from the top of which defenders could indulge in such horseplay as dropping stones, boiling pitch or quicklime onto unwelcome callers. The Duke of Clarence also added a tower, before taking up residence in London's Tower, and tackling rather more malmsey wine than he could manage. John Dudley was a one-time owner of the castle, but was careless enough to lose it, and his head, because he backed Lady Jane Grey rather than Mary Tudor – just another unfortunate gambler who met the last trump because he wrongly played his Queen. Charming relics include instruments of torture, together with Cromwell's death mask, and a rather poignant sight is that of the tiny suit of armour for the Earl of Leicester's son – he died aged three. The grounds were landscaped by Capability Brown, and the white marble 'Warwick Vase' (found, in fact, in Tivoli near Rome) is housed in a conservatory.

The town was extensively damaged by fire in 1694, but Georgian houses rose from the ashes to blend with the medieval remnants, and 'The St Mary's choir escaped' – one has a momentary picture of a stream of scorched choristers hot-footing it through a stained-glass window in the general direction of the Avon. St Mary's ten bells play a tune every three hours, anyone playing a wrong note no doubt being referred either to the ducking stool in the crypt, or the stocks in the Market Hall. These latter were mounted on wheels, and the unfortunate occupant had to drag himself into the railed courtyard to be locked in. Other attractions are the Lord Leycester Hospital used as ex-Servicemen's almshouses, the Doll Museum in the Elizabethan Oken's House, and the medieval tapestry map in the County Museum, no doubt with "Ye are here" tatted in with red silk.

Royal Leamington Spa

A spacious mid-Victorian spa with attractive buildings, some designed by J Cundall, a local worthy, and fine ornamental gardens named after Dr Jephson who managed to sell the idea of taking the 'waters'. These stem from natural saline springs, and feed the baths in the Pump Room. Their value in treating rheumatism was first recorded in the Middle Ages, and they so impressed Queen Victoria that she slapped the Royal prefix on the place in 1838. Visitors may either taste the waters from a fountain in the Pump Room Annexe, or whet their whistles at an outside tap (Matt chose the latter, and rather wished he hadn't), and some 50,000 water takers arrive each year. One of the features of the town comprises the colourful line in shop names, with examples such as 'Head Start' (hairdressers), 'The Go Between' (sandwich bar), and 'The Chair and Rocket' (pub).

Such were the jottings of Matt, and in his enthusiasm of this new hobby he started to give serious consideration to travelling the country preparing light-hearted articles of this kind that might give pleasure to readers, and even stimulate interest in following up the history of this country. However, Joyce brought him down to Earth by pointing out that this would be a full-time occupation in reading up all the background and developing a suitable professional style of presentation. After all, he had come a long way up the learning curve regarding technical writing, so he must expect to start at the bottom again for this new sort of writing. On reflection, Matt had to agree with her that it was really no more than a hobby, and a sort of minor escape valve from formal and disciplined writing, but was something he could always fall back on if he felt like it, as events were to prove.

Other events that stuck in their minds were the copious amounts of flotsam, largely plastic, that had to be dredged out of the weed hatch whenever they passed some of the canal-side factories; nevertheless the canal water didn't seem to have such a strong characteristic odour that they had noted during earlier visits. This surprised them a bit, since, unlike a river, the canal water is relatively static until the locks are opened to disgorge their contents downstream, and then gulp in another load from upstream. Even so, pollution from some chemical factories was rumoured to be sufficient for films to be developed in local rivers, and Matt recalled that some continental rivers had been impregnated with so much flammable waste as to constitute fire hazards! They also giggled at the sight of poor Dennis, who was helming at the time, going through a large lock in company with a boat helmed by a rather talkative lady, ready to recount her life story at the drop of a paddle. They could see Dennis's eyes becoming increasingly glazed as his ears were unceasingly bashed, eventually emerging suffering from 'lock jaw'.

Homecoming was a little melancholy but then, everything must come to an end, particularly good times. They all agreed to have built up wonderful memories, and felt remarkably fit after all those active days scrambling about the boat, and the countryside. The really sad time came when they took the Walker family to Heathrow, and bade them farewell, not knowing when their next merry meeting might be. However, in these days of rapid flight speeds and easier communications, continued contact with each other should not be too great a problem. Mum and Dad felt it most, of course, because they could not bank on still being around when the next meeting took place, so Dennis made a point of promising to arrange for plentiful supplies of photographs to show their progress at home, and particularly of the development of Harry. The family saw them off at Heathrow, but their sadness was not eased by the fact that the cause of the Comet crashes had not yet been determined. It took many months of painstaking research to locate the cause of the problem as fatigue near a window.

Chapter 14 – University (1955 – 60)

Following their canal holiday, there were still a few days free before starting his new job, so Matt took the opportunity to update his car. He found it surprisingly easy to arrange a part exchange for a recent *Vauxhall* saloon, and was exhilarated by the acceleration and comfort it provided, but was sorry to see the old *Morris* go, even though the front doors, hinged at the rear, had warped and so scooped in the cold outside air, making travel blankets essential. But then men do show affection for their mechanical 'toys', invariably classifying them as female.

Starting at the university was an exciting moment, and the first few weeks represented a steep learning curve for Matt, as he met all his new colleagues and familiarised himself with his office, lecture rooms and laboratories. There was no fuels laboratory, as such, and it was one of Matt's primary duties to set one up, so his early task was to list all the tests that he considered relevant, and then check the market for likely equipment suppliers. Within a few months he had organised the fitment of a laboratory with benches, sinks and supplies of gas and electricity, and this was followed by receiving and commissioning equipment for testing fuel density, distillation, calorific value, vapour pressure and a number of other properties. Then, with his assistant, Sid, he set about the task of preparing suitable instruction sheets ready for the students' laboratory periods. He felt that this was the most enjoyable job he had ever done, since the university showed so much confidence in his judgement on these matters, allowing him to use his own initiative to a great extent. Matt then sat down and drafted out the structure of his lecture course. (Details can be found in Appendix 3.)

The time came for the departure of the occupants of the cottage, so Matt, Joyce and the children moved in. There are always many things to attend to, however well previous occupants have looked after a place because, of course, peoples' needs and tastes differ, consequently it proved absorbing to arrange the furniture and ornaments to their own liking. Matt also fitted up an intercom with the house, so that instant contact could be made with Mum and Dad should the need arise. As before, Matt organised his workshop, in this case in the brick shed next to the garage, and Steve became a regular visitor, expressing interest in the names of the hand tools, and their use, so Matt encouraged him to shape out a shoe horn from a sheet of brass, and to inscribe 'Mum' on it. It was difficult to judge which of the two – Joyce or Steve – was the more proud!

As part of their house-warming process, they invited Brenda, Norman and Dulcie over from Bedford. Norman recounted how he had been torpedoed during a convoy from America, but was picked up later by another ship as a RAF plane appeared and drove the U-boat off. In fact, he said, it was quite noticeable that U-boat attacks were much reduced in the later stages of the war. It almost looked as if their plans were known beforehand! It had been a very tricky time, particularly in mid-ocean beyond the range of protective aircraft, and he remembered seeing some MACships, that is, merchant aircraft carriers, that were oil tankers with their decking modified to take an aircraft catapult. At times of greatest danger, the pilot would take off, locate any U-boats for attack, and then ditch into the sea, being picked up by the ship. All very dodgy – and expensive – but the blockade was eventually broken, helped by such imaginative and desperate techniques.

Norman also told them about his subsequent experiences as a rating officer. Inevitably some people disagreed with his assessment, and tended to show their displeasure in a physical way. One character even had a shotgun lying on his desk between the two of them but, fortunately, was not provoked enough to use it! Joyce and Brenda thoroughly enjoyed their talking over old times at the hospital, and Norman was very interested to see Matt's photos and jottings of their canal trip. As an ex-mariner, he would welcome again the feel of a moving deck under his feet, even if it was only a barge on a canal, and he vowed he would put the suggestion to Brenda for a similar holiday.

Matt often found his best ideas surfaced during the early morning, and while shaving one day, he suddenly thought that a problem that had arisen in car engines ought to be amenable to some fundamental research. The problem was that, after a long journey, when an engine is switched off there were times when the engine would continue to run, but in a very erratic manner, sometimes firing and sometimes not, and on occasions even going backwards! This was thought to be due to ignition caused by overheated surfaces of sparking plugs or combustion deposits. Matt had long ago realised that this was the reason why those precious Merlins of his were shut down by cutting off all fuel supply during slow running, and then switching off the magnetos afterwards – even though the plugs fired, the engine could not run without fuel. A sudden change of direction in rotation, of course, would wreck the gearing to the supercharger which ran at about ten times the speed of the engine! Why not, thought Matt, design a test rig based on an electrically-heated rod inserted axially in a quartz viewing tube, and pump a mixture of air and vaporised fuel through the tube, noting when ignition occurred. You would then have control of mixture strength and velocity for various fuel types, and could plot surface ignition temperature against contact time for conditions that are known and carefully controlled, unlike the situation in an engine. Later on, these rig results could be compared with those obtained from an actual engine.

Matt was excited by this idea, and drafted out a brief proposal. The Departmental Head was cautiously optimistic, and suggested that it might have applications to other situations such as fire safety. He went on to suggest that Matt prepare an application for a government research grant to support this project, which he would then countersign and submit for approval. Whilst awaiting the outcome, Matt continued to scour the literature for background knowledge of the theory involved, and of any comparable experimental work. Eventually there was much satisfaction all round when approval was forthcoming, and plans for the test rig were passed to the workshops.

The rig gradually developed into a bank of bottles of compressed air, and a boiler to vaporise the liquid fuel and inject it into the air stream leading to the concentric tube unit that Matt had designed. Vaporising the fuel proved to be the most difficult aspect because it tended to leave the boiler in surges, but increasing the free surface area of the liquid solved the problem. Safety measures were built in, and results were then obtained for individual hydrocarbons of known composition. The facility was therefore ready for some comprehensive research. 'Oh Fred, if only you were here with your Jungle Juices,' sighed Matt, 'We could try them all out.'

A phone call came through to the office while Matt was writing up his notes. "I hope you haven't forgotten that it's Steve's tenth birthday tomorrow," queried Joyce.
"Cripes," replied Matt. "I remembered to get him a card, but forgot to post it. I'll send it right now, and it should arrive in the morning. Thank you, Darling, for reminding me."

Ignition research had to stop at that moment until the all-important birthday card had been hastily consigned to the post box. Fortunately, the present of a small camera had already been secured, and hidden in the wardrobe. As an additional birthday treat, the family went to a film show and then a restaurant, giving them all an opportunity to relax. It was while Matt's brain was in neutral gear, as it were, that a sudden thought occurred to him. He waited until the next day, and then put the idea to Joyce while they were washing up.

"Look, my Dear, you know this report I'm supposed to write about the research I'm playing about with?"
"Yes, what about it?"
"Well, if I were to include some theory, and a bit about the applications in practice, and then tart it up a bit, I reckon it might qualify for a PhD thesis. After all, the work has to be original for a Doctor of Philosophy, and this rig is entirely my idea, and so is the boiler."
"I see. Would it mean a lot more work – for you I mean?"
"A bit. But then, I'd like to make a good job of the report, even if I didn't go for a higher degree, so with a bit more effort I could probably manage a thesis. The degree would certainly help in the academic world, even if nowhere else."
"I must leave the decision to you, Dear. You know what's best for us all, and I'll back you in whatever you decide."
"Bless you, my love, and thank you. I'll give it more thought and see how it pans out."

In the event, it did mean quite a lot more work, as Matt's deeper literature search revealed, but there was no evidence of any experiments exactly similar to those proposed by him. The one exciting event that arose during the tests was when the quartz tube cracked, and burning fuel spilt onto the bench. Matt did no more than reach for the extinguisher, and point it at the flames. Unfortunately, the extinguisher just did not work! After an agonising second, Matt was relieved to find Sid at his elbow complete with a serviceable extinguisher which did the necessary job. As Matt repeatedly said, engineers do like to have a back up!

After a review of the velocity and temperature variations in the boundary layers over his experimental heated surface, Matt reckoned he had achieved the beginnings of a balanced and potentially useful thesis. What was then required was a set of results from an actual piston engine, so the next phase was to insert heated surfaces within the cylinder of a test engine, and to correlate the results with those from the tube rig. This took some time to achieve but was eventually forthcoming, and so Matt registered as an external student of London University, and submitted the required number of bound copies of his thesis. It was then a case of awaiting any invitation for oral examination.

*****

The launch of the Russian satellite, Sputnik-1, caught the world by surprise, but added relevance to the rocket oxidant demonstration that Matt had been preparing for the approaching Open Day when visitors, particularly the families of the students, were invited to see the activities of the university. Matt's demonstration formed an interesting and attractive feature of the occasion, and he was delighted to note that Joyce, Mum and Dad were in his audience. His presentation then went as follows:

Ladies and Gentlemen,

Before I start this oxidant demonstration, may I illustrate a small problem that arose when refuelling an aircraft some years ago. Since aviation petrol is of particularly high quality, it is dyed green for recognition, whereas motor fuel in the Forces is coloured red. One day when an aircraft had been refuelled, the engineer flashed his torch into the tank opening to ensure that he had the correct fuel on board and saw to his horror that the fuel was red! He ordered the aircraft to be defuelled, and then noted that the colour was, in fact, green! What had happened was that the light from the engineer's torch had passed through the bulk of the fuel, reflected off the bottom of the tank, and passed through the fuel again to his eye. As a result, most of the colours in the torch light had been absorbed by the fuel, leaving only the red, which is the most penetrating." (Matt's thoughts went back to the approaching cruiser lights in Crick Tunnel.)

Sid here has made up this rig for us comprising two long vertical glass tubes containing green aviation petrol. We can see that one of the tubes is full, and the other contains only a few inches. Sid will now switch on this light underneath, and if we look into the mirror inclined above the tubes we will see that the shallow fuel still looks green whereas the deep fuel looks quite red. It just shows, we can't always go by looks.

Now for the rocket oxidants.
Most heat engines operate by burning a fuel with the oxygen contained in the atmosphere. But if we want to travel in space where there is no atmosphere, we have to carry our oxidant with us. The most convenient is oxygen itself, in liquid form to save tank volume, but we can also use oxygen-rich materials like hydrogen peroxide and nitric acid. The trouble with liquid oxygen is that it has a very low boiling point of -182°C, so it has to be contained in Thermos-like tanks to stop it boiling away, as I'll now demonstrate:

1. As I pour some liquid oxygen from this Thermos into this beaker of water, we can see the oxygen boiling vigorously as it extracts warmth from the water. Globules of oxygen can be seen forcing their way down into the water. This is because the vapour resulting from the boiling travels upwards giving a

downwards thrust on the globules so that they act like miniature jet engines.

2. Despite the very low temperature of the liquid oxygen, we can see that I can dip my hand into it quite safely, and that my fingers do not freeze or break off. This is because the warmth of my hand generates a glove of vaporised oxygen around it and insulates it from any more heat transfer. For a similar reason, a wooden stick can be used to stir molten steel, the vapour from the wood moisture acting as an insulator.

3. I can now demonstrate this even more dramatically by lifting up the beaker of liquid oxygen – and note that I am using a thick glove otherwise my hand would freeze to the beaker – and taking a small sip into my mouth, then breathing out through a lighted cigarette." (The cigarette burst into flames, and the audience into applause.)

4. Of course, oxygen systems in a rocket vehicle have to endure this low temperature continuously, and some materials harden up so much that they lose their flexibility and become useless. This piece of rubber tubing that has been standing in liquid oxygen for a few seconds has become so hard that I can smash it with a hammer. We therefore have to use special plastic materials for tubing and diaphragms.

5. In the home, we use hydrogen peroxide as a bleach, but that is at very low concentration in water. As a rocket oxidant, we need concentrations as high as 80% or more – we call it HTP for High Test Peroxide – and we also need some help to break it down into oxygen and water. Metallic silver acts as a catalyst for this purpose, and I have here a piece of copper gauze that has been coated with silver. When the gauze is dipped into the HTP, we can see that boiling takes place readily. Potassium permanganate acts in the same way.

6. Of course, once the oxygen is released from the HTP, it's available to burn with the fuel. It saves a lot of complicated ignition equipment if the fuel and the oxidant are able to ignite spontaneously when they meet. We say they are 'Hypergolic'. For example, hydrazine hydrate fuel and HTP oxidant are a hypergolic pair. But it is vital that they ignite very rapidly on meeting, otherwise the rocket chamber will be full when ignition eventually does take place, and this will result in a catastrophic explosion.

When I pipette some HTP into this dish of hydrazine hydrate, we can see that it does, in fact, ignite very readily. But we have no chance of measuring the length of the time delay between meeting and igniting using this method. However, we can do this by using the twin jet rig here. We can see that this has two glass tube jets lying horizontally in the same plane. When Sid switches on the HTP, and then the hydrazine hydrate, we can see that the two horizontal streams start to fall under gravity. Because they are inclined to each other, they first meet at a fixed point, and after mixing they ignite at a second fixed point. All we have to do is use this mounted telescope to measure the depth of fall from the jets to the point of meeting, and then to the point of ignition. It's a straightforward matter to convert these distances into times, and take the differences in them as the ignition delay. In the demonstration we see here, I would estimate the delay to be about 60 thousands of a second. Ingenious rig, isn't it?

7. Hydrogen peroxide will also decompose slowly when in contact with organic materials like flesh, paper or cloth. In this bucket is some screwed up newspaper, and I will now pour some hydrogen peroxide onto my hand, then into the bucket. We can see that the paper catches fire almost immediately, whereas it takes a few seconds for the white blotches of oxidation to appear on my skin. Of course, if the peroxide has splashed onto my lab coat, this would probably catch fire in about five minutes. So now I will take the coat off and hang it

on this peg here – and I suggest we all leave the lab now before there is any fire.

Thank you very much, everybody, for your attention. And thank you, Sid.

Matt had enjoyed himself hugely, and was pleased that the family had also found it interesting. He was even more pleased when an industrial film unit arrived later to record the whole demonstration, and present him with a copy. (Many years later he was to transfer this to a compact disc.) He made a mental note that he was building up enough material to attempt a book on spaceflight propulsion, including all the exotic electrical, ionic and nuclear systems as well as the more conventional fuel and oxidant methods, but he had to improve his fuel book first.

* * * * *

Then came sickening news of the two trains crashing under one of the bridges at Lewisham in thick fog, with the loss of 92 passengers, some of whom were friends of Matt and Joyce during their days in Brockley. One of the poignant reports was of the remains of a little lad with his drum beside him that was his gift for Christmas. Matt's reaction was that the steam locomotive must now be replaced with a vehicle giving direct vision ahead for the driver, and perhaps some automatic braking system operated by the signals, in order to help avoid such tragedies.

One afternoon, Steve announced that he wanted to go into the woods and photograph the bracken in the gold light of the Sun. Matt accompanied him, and listened carefully while Steve drew his attention to some of the plant life appearing within the leaf carpet. Suddenly, Steve raised a question that caught Matt by surprise. "Dad, the boys at school say that girls bleed every month. Is that right?"

Matt grabbed at the chance for some 'boy talk', and was delighted that Steve was showing sufficient confidence in him to enquire openly without embarrassment, rather than relying on the 'behind-the-bike-shed' grapevine for intimate details of this kind. Matt recounted the routine of an egg being produced within the female, and waiting to be fertilized by the male sperm. When this does not occur, it is no longer wanted and so is flushed out by a flow of blood, in a process known as menstruation. Together they marvelled at this apparent natural miracle of two sets of genes being fused to determine not only the physical and mental characteristics at birth, but also the inbuilt changes that arise at different instances during the years of life ahead.

"What I still don't understand," conceded Matt, "is why menstruation happens so regularly every month, and whether it's controlled by the Moon in some way. What I do remember is that a lathe operator in Woolwich advised his boss that his lathe would become slightly inaccurate at each full Moon. Not surprisingly, this was treated as ridiculous, but it was shown later to be true, and it was considered to be due to the Moon attracting the water table which meant that the ground swelled up slightly and distorted the lathe, even though it was made of a heavy casting."

When the two reached home, both felt a flow of satisfaction: Steve in that he now felt he could ask his Dad anything at all, and would get an honest reply, and Matt that his son respected him enough to seek his advice rather than go elsewhere. He also reckoned that, at this rate, Steve would soon be able to advise him on a host of topics in natural history.

The expected invitation to be interviewed arrived from London University, and the family drove to London and waited in the car while Matt presented himself to the examining panel. Once again, Matt found the experience thoroughly interesting, albeit a little nerve-wracking, with knowledgeable examiners asking direct and pertinent questions. It was clear that they were all on the same wavelength with regard to the subject of spontaneous ignition, hence it was no surprise when, following the usual nail-biting wait in a side room, Matt was invited back and congratulated on his success. Once again, also, he was advised that he could not add the qualifying letters, in this case PhD, after his name until he had received written confirmation from the University authorities.

The family did not have to ask the result of the interview when Matt returned to the car, as it was written all over his face, so after finding a parking spot for the evening, they set off for a special meal, followed by a family-type film. The voices of Mum and Dad on the end of the phone were buzzing with excitement, and they promised to advise everyone else in the family, the neighbours, and anyone else who cared to listen!

Matt was particularly impressed with the arrival of the hovercraft since, to him, this meant such a practical compromise between high speed without the problems associated with fuel handling and combustion at high altitude. He was even more impressed when it crossed the Channel the following year. Also exciting was the opening of the M1 motorway, making it so much easier to drive to London or, when the occasion demanded, to Leeds and other northern locations. Most drivers seemed to learn easily how to merge into the main traffic stream from a junction. There was no upper speed limit in these early days and, just for the experience, Matt managed to 'do a ton' during a quiet stretch.

By this time, Matt had managed to revise his Fuels book to his complete satisfaction, and to get it published. Even though he did not expect to receive much money from royalties, it would prove to be a valuable source of worked-out arguments and carefully prepared illustrations for his lecture courses.

* * * * *

It was the day Joyce went to take tea with a friend in a village shop that the alarm bells started to ring. It had been arranged for Matt to collect her at five thirty to bring her home, but the phone in his office rang at four, with a call from a somewhat worried Joyce.
"Matt, Dear. Could you come and pick me up now – if it's not too much trouble?"
"Certainly, my love, but what's the problem?
"Well, you see," somewhat hesitantly, "I've just coughed up some blood, so I'll need to see the doctor as soon as I can."

Matt felt his heart lurch, and his stomach tighten. What the blazes is all this about? He thought he'd noticed that she'd been a bit under the weather recently. He must get to her at once. And get her to the doctor, too.

Matt drove purposefully to the village and parked outside the shop. Joyce looked cheerful enough, if a little pale, but Matt wasted no time in arranging for an appointment with their GP, who reported that there was probably nothing to worry about, but that he was arranging for her to be X-rayed at the hospital. Various tests followed, and it transpired that Joyce was suffering from tuberculosis that had caused a hole to develop in her right lung. The local hospital then transferred her to a specialist who advised that, although the disease itself had been halted, the recommended treatment for the lung damage was to remove several ribs from the back so that the body compressed the lung and stopped the hole from developing any further. With her nursing background, Joyce could understand the reasoning behind this very well, and accepted it philosophically, but Matt was shattered. The thought of Joyce's lovely body being carved up and disfigured was something that he could hardly accept. As an engineer, he wondered why they couldn't inject some porous plastic into the lung to fill the void, but of course he had no expertise whatever in medicine.

Matt organised all his spare time to care for Joyce, the children and the house, and his colleagues were sympathetically assisting with his academic duties. The situation was eased significantly when it appeared that Dorothy, Matt's second cousin, was having to leave her flat because of the end of her tenancy contract. Since she had nowhere else to live, Matt's parents invited her to stay with them at Brickhill. This she was very grateful to do, but as soon as she arrived and realised the situation in the cottage, she offered to move there instead and give what help she could to Matt and Joyce. The timing couldn't have been better, and Aunty Dot became one of the cottage team. Dorothy was grateful to be living with people again, particularly of her own family, but nowhere near as grateful as was Matt who, truth to tell, was no great shakes at cooking, and knew even less about changing the bedclothes once in a while, although he had mastered the washing machine, and even typed out simple directions for operating it in case anyone else should be involved.

Aunty Dot also was a remnant of that band of people who had been 'in service with the gentry', and then forced to seek other openings when that style of living steadily eroded, accelerated by the war. Dot had been employed as an assistant cook, and could well remember seeing stacks of jelly moulds in the older kitchens, testimonies of the days when tooth decay went untreated, and people were forced to eat their victuals in aspic and other jellied form in order to be able to swallow. Some people, it was rumoured, even had to wear wooden teeth! With the outbreak of war, Dot had volunteered for the ATS, and served in the cookhouse. On her discharge report she was described as a 'Good plain cook' and, although she did not possess the culinary skills of Joyce, she could be relied upon for providing nourishing square meals for the family, and her bread-and-butter puddings sent the children into raptures. She also idolised Toby, and he her. One further bonus was that Joyce had another woman to talk to.

One of Joyce's frequent visitors was Brenda, who was deeply concerned with her friend's health, and anxious to give what support she could. Matt was most grateful for this, particularly when Joyce had to attend hospital for her surgery. It was an agonising wait for the phone call afterwards, but Matt did his best to concentrate on revising his lecture notes at the time – not that this was at all urgent, but it kept him occupied. When the call came through that she was back in the ward, and that he could visit, he lost no time in getting there.

As Matt approached the ward, he wondered what line best to take with her. Should he seek details of the operation, or keep off the subject of surgery altogether? Whatever he did, he wanted to bring her strength, love and support. He needn't have worried. On the notice board outside the ward door were pinned dozens of cartoons cut from newspapers and magazines depicting jokes about hospitals in general and operating theatres in particular. It was clearly the patients' way of cocking a snook at their condition, and accepting their situation with courageous and philosophical gusto. Much heartened and relieved, Matt stepped inside and was directed to Joyce's bed. She looked somewhat fragile, but otherwise cheerful, and wanted to know all about the children, Dot and his parents. Matt instantly cottoned on to his new role of the news vendor for Joyce's benefit, and very shortly after, he bought a small notebook in which to jot down items of family interest that might appeal to her. Whilst there, he arranged for a telephone to be brought to her bedside, and for Joyce to have a brief word with Brenda. She then wanted to call her in-laws, but Matt thought that she had had enough for one day, and promised to convey her wishes to his parents on her behalf.

The next few months were trying for all concerned, but eased by the fact that Joyce continued to make steady progress. Matt's journeys to and from the hospital were so regular that it came as a great relief all round when it was agreed that Joyce could return home as long as she had someone (Matt plus Dot) to nurse her. Matt's parents took the children out for the day, and Matt drove to the hospital with a heart lighter than it had been for months. He helped Joyce gently into the car, and bore her home with a feeling of relief.

For Joyce, there was one big problem. Although she knew that Matt loved her deeply, she was well aware that, as a woman, she had lost some of that physical beauty that she perceived to be so important to a husband. It was with some trepidation, therefore, that as she prepared for bed at home, she stripped off topless and showed herself to Matt. "Well, this is how I am now, Darling. Is it too bad?"

Matt looked fixedly at the great valley carved into her back, with its scar still pink from surgery, and then to the shrunken right breast that it had created. Thankfully, that favourite left breast of hers that had always given him so much delight and comfort, was just as gorgeous as ever. He went over to her, and took her gently into his arms.

"My lovely, lovely Joyce. You are as beautiful as you've always been to me, and I love you even more, if that's possible. Also," he added with a smile, "you turn me on just as always, so hurry up and get your strength back so that you can study the ceiling again when we're in bed."

Joyce couldn't speak. She just sobbed quietly as relief surged over her. Her man still wanted her, and that was all that mattered just then.

Chapter 15 – Oz (1960 – 65)

Although Joyce had lost the use of part of her right lung, the body tends to compensate, and she was also wise enough to limit her lifestyle accordingly, so things settled down in the Gregg family to almost the same pattern as before. The main drawback was that Joyce was prone to pick up colds more easily, and finding it difficult to be rid of them, so congestion became a bit of a problem. Also they found that their social life was ruined by people smoking because, even if they were at the far end of the room, it quickly affected Joyce's breathing. It was while they were enjoying the reading of a letter from Dennis and Val that Matt had this idea of getting Joyce to a warm country for a while to 'dry her out' and help her lungs to strengthen. When a notice appeared in one of his institution journals that senior staff were required for a four-year period to assist an Australian university college to gain its autonomy from its parent university, Matt's imagination was sparked, and he discussed the idea with Joyce and then his parents.

It was rather a pity that Dennis and Val lived in South Australia, whereas the position advertised would be in New South Wales but then, if they went, they would at least all be in the same country, vast though it is. The whole family agreed that a warm climate could be beneficial for Joyce, and it wouldn't do Matt's career any harm to gain some more senior experience for a while. Even though the present job would probably not be held open, it seemed worth the risk to have a four-year break overseas, and take his chance for a job afterwards, bearing in mind the additional experience he would have.

Buoyed by this thinking, Matt prepared an application for the post, and an interview was duly arranged at Australia House, Strand. This is when the new M1 motorway was a great help in reaching London more easily and quickly. Although Matt's qualifications and CV virtually spoke for themselves, the board was particularly interested to know about his personality, ability to communicate and the balance and maturity of his outlook on life. Matt's questions about the college facilities and local housing situation were answered satisfactorily, plus the fact that, if appointed, their passage and removal costs would be covered by the college. It was no great surprise, therefore, when an offer arrived for Matt to take up the position of Associate Professor to head up the Department of Mechanical Engineering for a period of four years, with a possible extension.

There followed an enormously exciting period for everyone, with lists of 'things to take' being hastily prepared, amended, scrapped and rewritten. Aunty Dot moved over to the house with Mum, Dad and Toby, and the cottage was prepared for re-letting. The university expressed regret at Matt's departure, and gave a memorable farewell dinner. In the lounge afterwards, Matt was delighted and deeply touched to find that, with Joyce's connivance, his colleagues had been to his house and checked through all his family photos, so that they could present a 'This is your life' programme on him right from his lying as a naked baby on a rug to the present day. To think they would bother to do all this, meant so much to Matt. When he finally went along to the workshop to say farewell to 'the lads', their banging on their benches with their hammers rendered him almost speechless with emotion. He would certainly miss them all, and vowed to see them again when he returned, wherever his job may be.

Departure from the railway station was also an emotional experience, and Matt could not avoid seeing the teardrop in Dad's eye. Having lost one son, Dad hoped against hope that he would still be around to see the second son return with his family. At Tilbury, they went through the usual facilities and found themselves on board the *P & O Strathmore*, one of the old-time liners complete with polished mahogany and Goanese deck hands. The Bay of Biscay lived up to its reputation, but the Suez Canal, of course, gave them complete calm. Stops at places like Suez and Singapore proved very educational for them all, particularly when they went ashore at Aden to find, even at two o'clock in the morning, rows and rows of shops stuffed with cut price radios, cameras, watches and a host of other goodies – and yet there were people sleeping in the gutters, and hungry cats prowling about for food. It was a once-in-a-lifetime opportunity to invest in a projector and screen to complement Matt's movie camera, but there was a clash of conscience about whether to spend on such items when poverty was so rife. However, Matt consoled himself with the thought that if he spent money here, there was a chance that it would filter down to the needy, whereas if he didn't, there was no chance at all. So he spent, and returned to the ship laden with items for the family and himself. The ship then departed, bristling with the antennae from a thousand newly-purchased transistors; 'Fruits' as one wag put it 'from the Garden of Aden'.

The warm tropical nights at sea proved to be just like the paperback descriptions, with scented air and jewels of stars set in a velvet sky, so low that you felt you could reach up and touch them. In fact, long before Ceylon appeared over the horizon, the air was filled with hints of spices and other exotic odours, described by some other wag as 'Eau de Colombo'. Once ashore, the Greggs visited the zoo and a number of Buddhist temples. Later on they saw a snake charmer for the first time, and Steve was quite happy to let the creature be draped over his shoulders. Joyce and Avril were nowhere near as sanguine and, truth to tell, Matt wasn't too keen either. He had no doubt that its teeth or poison sacs had been removed, but even so – Yuk!

At the next portstops they found the flogwallahs were active, crowding their bumboats round the ship and dock and touting their wares of electronic goods and confections. There was a remarkable show of trust here because the flogwallahs would permit a potential customer to raise the item for sale up the side of the ship in a roped basket and, if accepted, lower the basket containing the appropriate money. The system seemed to work very efficiently, but if the flogwallah was tardy in removing the rope when the ship was ready to sail, he ran the risk of it being cut by the ship's crew, irrespective of what the basket contained at the time. A magician had come aboard and, to the repeated comment of, "Is quickly coming," produced what looked like day-old chicks apparently from space. No doubt, large hands and voluminous sleeves played some part.

Their first sight of Australia was of a strip of sandy-looking soil on the horizon, but they were much impressed with the scenery when they reached Fremantle and took a coach trip to the beautiful city of Perth, where they saw black swans. After sailing through the Bight, they disembarked at Melbourne. Although the original plan had been for landfall at Sydney, an engine repair to the *Strathmore* had led to a change of itinerary, and they were to continue their journey by rail. Matt learned something that, on reflection, he felt he should have guessed. It is general knowledge that the rail gauges differ between states in Australia – broad gauge in Victoria, standard in New South Wales, and narrower in Queensland – and Matt wondered how the Aussies could tackle the monumental task of country-wide standardisation. But they didn't have to. The Australian solution was to build standard-gauge rail links between state capital cities only, and then transfer passengers and goods to local lines as required.

Matt had promised himself that he would visit that lady in Melbourne to thank her personally for her kindness in sending the groceries during the war but, sadly, her address label had been lost when the house was bombed.

The train journey was quite adventurous in its way, because the carriages rocked noticeably at speed, and the standard joke was that, if not careful enough, your spoon of soup could finish up in the next person's mouth! Also, Matt was puzzled when the train slowed and, with no words spoken, most of the passengers got up and threw their newspapers out of the windows. It transpired that the slowdowns were caused by men working on the line, and, in such a large country, this was the only way they could keep up-to-date with the news and racing results! (Matt was to learn that, on the day of the Melbourne Cup, just about everything stopped throughout the country!)

Matt was impressed to find the Principal of the college waiting at Sydney station to accompany them to their ultimate destination, but was alarmed when a whole crowd of cheering students burst through the barrier and came charging towards them. Surely he would not have to go through a welcome of this magnitude! However, all was revealed when the students surged past him to the back of the train where some American pop celebrity was alighting.

Their first accommodation was in a local hotel, which is where they were delighted to entertain Dennis and Val who had driven all the way from Adelaide. Meeting members of the family like that was, of course, a great fillip when arriving in a new country not knowing anyone. Dennis and Val were able to give them many tips about successful living in Australia, and also taught them some of the 'Strine' language.

Matt and Joyce decided to buy a property rather than rent, just in case they could persuade the family back home to join them, so house-hunting became a major priority. They were intrigued to find the conventional house in Australia to be a bungalow of wooden construction (very hard wood indeed, as Matt later discovered when he tried to hammer in some nails), mounted on brick blocks, allowing cooling breezes below the floor… and also providing habitats for poisonous red back and funnel web spiders! Surmounting each house at considerable height was an H-shaped TV aerial. The Greggs eventually settled on one of these houses on a block of land in a suburb, and it wasn't long before Steve collected a tick boring into his head, leading to the discovery that it was no good trying to pull the confounded thing out since you could finish, literally, with half a tick. The trick, apparently, was to apply a drop of fluid, like paraffin, to persuade the creature to go into reverse.

Next in importance was the purchase of a car, and Matt rapidly homed in on a utility type that would carry their baggage as well as themselves. Petrol at the equivalent of 2s.10d a gallon was a joy. The first impression of the city included shops with fixed awnings covering the pavements, no doubt as shelters from the strong sunshine (and the equally strong rainfall when it arrived!). Matt then noted that the catering front seemed to be dominated by sponge cakes, lamingtons, tomato sauce and passion fruit, plus plentiful lamb chops. Joyce, on the other hand, was introduced to squash as a vegetable, something she had never even heard of. The impression growing in their minds was that the Australian culture seemed to have adopted much of English tradition but spiced it up with American innovation. The US spelling of 'favor', 'color' and 'harbor', for instance, was widely used, and the ideas of the supermarket and outside gas and electricity meters seemed to have caught on much more quickly than in the UK.

Another Americanism was the casually efficient means of the delivery of mail and newspapers. Matt had searched in vain for a letter box slit in the front door when they moved in, but then realised that the custom was to have a letter box located on the gatepost close to the roadway so that the cycle-borne postman could effect delivery without halting his momentum. Registered mail, of course, was not so easily dispensed, and the 'postie' was forced to bring his machine to rest in order to blow his whistle as he summoned the recipient to emerge and sign. Newspapers, on the other hand, were usually rolled up and banded at the shop beforehand, and the deliverer, who had no van door to contend with, merely adopted strong, well-directed sweeps of the arm to left and right, with timed release of the fingers. The householder then had the intriguing task of 'hunting the paper' in the front-garden vegetation. Garbage men worked on a similar principle of rapidly handling wheelie bins, neatly arranged by the householder in military array by the roadside, the suburban roads themselves having tar-sealed crowns but unmade edges. Since glass bottles had not yet been replaced by plastic receptacles, milk deliveries were more conventional, but a bottle-for-bottle system was practised rather than leaving out little notes for the 'milko' to specify the required number of pints. In the event of any query, questions could be asked from outside since sound travels quite well through weatherboarding.

A truck pulled up at the college delivering all their tea chests that had also travelled by sea, and Matt started ferrying them to their new home, at the rate of two crates per journey. It was almost like Christmas when opening the crates, because they had no idea what was in them, and things turned up in no logical order whatsoever. However, it was all inspected, dusted and located where required. Only one small piece, thankfully insignificant, appeared to have been broken during the voyage. So, quite shortly, they found themselves all set up and ready to go.

One of the inconsequential things that reminded them that they were no longer in England was that every time they saw a map of the world, Australia featured in the centre. However, this meant that England appeared twice, being included on both the western and eastern edges. But the first real shock Matt had was that the water rates were about ten times those in England. It brought home to him dramatically how vital this commodity is in a hot country, whereas he had tended to be seduced into thinking that oil was the most important fluid in the world. Also, when the time came for house painting, he realised that that was exactly what was meant. All the wooden weatherboards of the house had to be painted, not just the doors and window frames as back home.

Matt and Joyce were pleased to find the friendliness extended to them by colleagues and neighbours, and the children settled in well at school. Matt found it refreshing that Australians do not accept people on their titles or qualifications alone, but rather on their personalities, i.e. 'is he a 'good bloke?' The womenfolk raved over Avril's peaches-and-cream complexion – it appeared that the strong sunshine tended to dry out and crease the ladies' necks unless they wore broad-rimmed hats.

Whenever time permitted, the family made a point of touring to see as much of the countryside as possible. A kind offer from the water authorities saw them enjoying a boat trip on the headwaters of Glenbawn Dam, an experience that registered deeply with Matt, and they then visited the scenic Blue Mountains, enjoying the magnificent peaks of the Three Sisters and exploring the spooky wonderland of the Jenolan Caves. Matt was given to understand that the blue coloration of distant trees was due to vaporisation of the natural oils from the leaves – hence Nature does a little polluting of her own! The scenery was stunning in a harsh, brilliantly coloured way, somewhat different from the more gentle landscapes of home.

One thing that intrigued the Greggs was that they had to stop at the borders between states, and either consume or destroy any fruit they were carrying. This became understandable when they were told about the potential ravages of orchards by fruit flies, but they could not imagine having to go through a similar routine when crossing from, say, Sussex to Kent. They were much impressed by the foresight shown in the provision at numerous tourist spots of brick-built barbecue stoves, with fuel wood stored nearby, and also automatic machines for dispensing ice, either as a block or chipped, for one's picnic box on the insertion of a two-shilling piece.

By this time, Matt felt he was in a position to correlate all these impressions into one list that he could send back home, and this appeared as follows:

The Greggs' First Impressions of Australia (1960)

We are delighted with:
    the sunshine and open-air life,
    the attractive detached houses, and open gardens,
    the price of petrol, coal and wines,
    the gaily-painted seats on stations and at roadsides,
    the tasty, widely available sandwiches, and chilled beer,
    kookaburras,
    the simplicity of the income tax system.

We are impressed with:
    Australians' national pride,
    the magnificent scenery,
    the surf and surf riders,
    the atmosphere of expansion,
    the high level of salaries and wages,
    the sharp rainfall and tropical thunderstorms,
    the gigantic proportions of tomatoes, butterflies, ants, cockroaches, refrigerators, TV sets, and the storm water drains,
    the practice of moving an entire house on a massive trailer by road.

We are intrigued with:
    fly screens and sun blinds,
    rotary clothes lines,
    out-of-doors location of gas and electricity meters,
    mailboxes on the gatepost, and whistling postmen,
    the practice of walking barefooted,
    the lizards,
    the wide use of the illuminated mechanical signalling hand on heavy vehicles,
    the tea cosies,
    pay-as-you-enter buses,
    children in night clothes on evening walks,
    drive-in cinemas.

We are surprised by:
> the cold, probing, westerly winds in winter,
> the limited use of sunglasses,
> the deafening noise of crickets in the evening,
> chickens crowing at night,
> the high prices of houses, cars, TV sets and some foods, especially sweets.

We dislike:
> the state of the suburban road surfaces,
> the road lighting, and the regulation to use full-beam headlights continuously at night,
> the lack of 'cats eyes' for night driving,
> the hazard of the traffic dome ('dumb cop' or 'fried egg') in the middle of the road,
> the trend of rising prices,
> the litter and broken glass,
> bloodsucking bush leeches, and flesh-boring ticks,
> compulsory voting,
> passion fruit pips in ice cream and fruit salad.

For their own purposes, Matt also devised a list of Australian expressions, with their English equivalents, and here he was helped greatly by Steve and Avril who brought home further examples they had encountered at school.

The Greggs' Australian-English Dictionary (1960)

Beaut (pronounced Bute) – Beautiful, lovely, smashing
Block (of land) – Plot
Blue – Row, barney, bust-up
Chooks – Chickens
Crook – Unserviceable, ill, useless
Drongo – Clot, idiot
Glory box – 'Bottom drawer'
Good on you, cobber – Nice work, mate
Grog – Drink, liquor, booze
Hooray – Goodbye

Lay by – Deposit to secure a purchase
Over the fence – beyond a joke
Port – Attaché case
Sedan – Saloon car
She's apples, she's jake – That's all right
Shower – Hen party to provide a potential bride with gifts
Shrewdy – Smooth operator, wide boy, spiv
Smoko – Elevenses, coffee break, morning tea
Stroller – Child's push chair
Subway – Bridge with low clearance
Tucker – Food, grub, a meal
Useful – Handyman

A minor domestic crisis arose early one morning when Matt had just entered the bathroom for the daily ablutions. He was called into the lounge by Joyce to find a somewhat tearful and defiant Avril. It appeared that Avril was seeking permission to have her ears pierced so that she could wear earrings on special occasions, but was sure that Dad would object. As a woman, of course, Joyce recognised the wishes of a growing girl, but to Matt it seemed entirely ridiculous. As we have seen, our Matt had no truck with personal adornment. He recognised that a young blade would be attracted by the hairstyle and figure of a girl if seen from the back, and her face as well if seen from the side or front. He always found Joyce to look absolutely lovely whether or not she was sporting any make-up, although he grudgingly admitted that she did look extra special when wearing those pearl earrings he had given her. He also thought that no young man, surely, would bother to look at the rings on a girl's fingers, unless his designs on her were financial, and he could not understand why girls put so much importance on rings, and handed them down so seriously throughout the family. What mattered to him was, what was in the brain, and how it was used.

However, Joyce explained patiently how a girl feels about her approaching womanhood, and the peer pressure that arises at these times, and Matt then recalled Mum's comment that, 'You might as well please them, as tease them!' So Matt concluded that, although he didn't really understand about the need for women to deck themselves with pieces of rock or metal, however rare and therefore expensive, he wanted everyone to be happy so agreed reluctantly, to much relief and joy all round. Suppressing his irritation, he then requested permission to return to the bathroom where, he explained, he needed to apply a damp flannel to several important little places. His humour was not restored by the fact that he recognised the inexorable progression of his baldness, but he gained some comfort from the fact that Steve was showing an increasing fascination with all things natural, and gaining quite an extensive insight into the flora and fauna of this new country. Steve was also showing a fair amount of skill in using the camera given him as a birthday present.

In fact, resulting from his interest in the natural world, Steve had received from a school friend a reptile that was known in Australia as a turtle, but would probably have been called a tortoise in England. This specimen was of the snake-necked variety, with his bulk following at least five inches behind his head as he stomped about the garden. When disturbed, he retracted everything, tucking his neck into some hidden recess near his right armpit. The hole drilled in the rear of his shell showed that he had been tied to some fixed point in the donor's garden. Matt and Steve enjoyed its company, and photographed it from every angle, but then noticed that the unfortunate beast was straining to be free from the string so that it could proceed to the swampy area at the bottom of the garden. Steve's eyes met Matt's and, with no words spoken, they both knew that they could not rest while this creature was under stress, so they untied the string, carried the turtle to the swamp, and watched as it cheerfully paddled off in the general direction of the river. They both felt contentment as they walked back indoors for morning tea, with the word 'Freedom' in the forefront of their minds. Although Matt didn't know it, Steve felt at that moment as much affection for his dad as Matt had always felt for his own father.

* * * * *

Matt drew his staff together in his office to survey the lecture programmes for the coming term. Since Matt had had experience of aero and space propulsion, it was agreed that he would cover the subject of advanced energy systems. This suited him because he would need to deal with some material about electrostatic systems that, so far, he had had no time to study in depth, and would also give him the incentive to complete that book on spaceflight propulsion that had been in the back of his mind.

As an academic of professorial rank, Matt was privileged to be invited to see the Snowy Mountains scheme near the capital city of Canberra – a vast endeavour centred in an area of snow larger than Switzerland. What impressed Matt the most was when his party was taken into a massive underground power station to see the water turbines driving the generators, and told that there were no human operators present at all – control was maintained from another power station many miles away! Matt was also invited to visit an oil drilling rig way out in the bush, where he was presented with a drill bit to take home. He was delighted to see kangaroos jumping about free, rather than behind zoo bars, and was intrigued to stay overnight in a nearby township. Here, the source of electricity was a large single-cylinder diesel engine giving regular one-second thumps that echoed around the town all day, *and all night*. He found it difficult to sleep, although nobody else seemed to. He learnt later that when the diesel engine was replaced by a relatively quiet gas turbine, the locals could not settle at night because of the lack of their familiar background noise!

When Matt returned home, Joyce showed him a letter from Brenda to the effect that Norman had taken an examination, and been offered a better position in an office in Norwich. They had therefore moved, but looked forward to meeting Joyce and family as soon as possible when they returned to England.

Matt and his staff had been following closely the developments in spaceflight with Yuri Gargarin of Russia, and then John Glenn of America. Matt was able to expand his notes on spaceflight propulsion, and introduce a course of lectures on the subject. He was also thrilled by the announcement that a supersonic aircraft, Concorde, was to be built jointly by Britain and France.

Now that the revised college courses were settling down, Matt and his colleagues had time to consider plans for a site for the expected new university, presuming that it would gain its autonomy. An area of bushland had been earmarked (it was rumoured to be a favourite haunt of the followers of an illegal 'two-up' gambling school), and it proved an absorbing task to outline the suggested relative locations of the various buildings in order to submit to the architects.

A sense of shock travelled around the world with the news of the assassination of President John F Kennedy in Dallas in November 1963, and then of his alleged assailant, Lee Harvey Oswald. Inevitably, political leaders will attract enemies but surely, concluded Matt, assassination can never be the answer.

A deeply sad moment arrived with the death of Sir Winston Churchill two years later. Mourning was widespread, but the moment that brought tears to all the Greggs' eyes was when the coffin was carried by launch along the Thames, and the dockside cranes all bowed as a sign of respect. Dockers are not considered to be the most sensitive members of society, so this gesture was hugely significant. Matt and Joyce only hoped that young people who had been spared the horrors of World War II would come to respect the man who had saved us from defeat.

Australia was so full of new experiences that time had flown, and Matt took the opportunity to jot down some of the impressions he had gained during their stay, and what he would be able to tell the folks back home. He listed his thoughts on the following lines:

1. Mateship

Australian men seemed to put a premium on their relationship with a 'mate', usually at work, and would stick by him through thick and thin even if, and particularly if, the mate was in trouble of some kind, including the law! This no doubt helped them through many difficult times of sickness, unemployment or war, so it was no surprise to find Ex-Servicemen's Clubs flourishing throughout the country. In fact, because of the readiness of the Aussie to try his luck, one-arm-bandit gambling machines in these clubs generated levels of income that were said to be embarrassing. It seemed that many an Aussie female was keen to indulge, as well. One of the funniest sights ever witnessed by Matt was in the local bank where a portly Australian in floral shirt and khaki shorts was loading his pockets with 'two-bob bits' ready to feed the machines. Eventually, the weight of the coins became such that the shorts slid gracefully to the floor, revealing brief pants delicately emblazoned with a matching floral design. Even though Australian men tend to be macho, they don't seem to enjoy much in the way of exposure, and acute embarrassment followed.

Matt was somewhat disturbed to find that, although the celebration of Anzac Day was observed widely without fail, some of the young were tending to think that this was just an excuse for heavy drinking. But, thought Matt, the young must be free to question and, hopefully in this case, come to discover the bonds that grew between servicemen involved in bloody battles between periods of mind-numbing boredom all those years ago and, sadly, even up to the present.

2. Police

Matt had memories of the local 'bobby' at home who would remonstrate if he caught youngsters scrumping, but proved to be a good and respected friend eventually, especially at times of lost children or accidents. However, he'd heard about the terrible conditions met when the convicts were deported to Australia, even for such a trivial sin as stealing a loaf of bread to feed a hungry family, and how harshly they were treated by the police of the day. Some convicts even became suicidal and jumped into the sea.

It seemed to Matt that these memories had been handed down, and that the populace regarded the police – who all carried guns – with thinly masked suspicion. This was underlined when Matt courteously halted his car at the roadside to give an approaching police car priority at a level crossing. The police car drew up beside him, and the officer wound down the window. 'Blimey,' thought Matt, 'I wonder what law I've broken now – perhaps they'll be lenient when they know that I'm a newcomer.' But the policeman merely said "Thank you" in a slightly surprised voice, and then drove off. Clearly he did not normally encounter such thoughtfulness, trivial though it was.

3. Dogs and Cats

When driving through the townships, Matt was surprised to find that a number of dogs would routinely chase his car, snapping at the wheels, the owners clearly not taking the trouble to train them otherwise. It turned out that this was not just peculiar to the local scene – apparently it happened country-wide. However the worst offender to Matt's mind, got his come-uppance when he hit his snout on the rear tyre, and retired with a look that could only be described as 'old fashioned'.

Matt was dismayed to find one day that a small terrier had been dumped on his veranda, necessitating a journey to the local animal hostel. It seemed that this was not an unusual occurrence. But Matt queried why the owners didn't take the animal there in the first place? One of the most endearing moments for him was the sight of a mongrel trotting along the high street with a huge juicy bone in his mouth – nothing peculiar in that, it may seem, but this dog also had a second juicy bone tucked in his collar. That must be one happy dog, concluded Matt, and one obviously on extremely friendly terms with his butcher! As an engineer, of course, Matt was nurtured on the wisdom of always carrying a spare.

There were a number of feral cats loafing about the college site, which were mainly untroubled by humans. With Matt's fondness for animals, he could not restrain from stroking the odd cat if he had the chance, and hoped they would help to keep down the army of huge brown cockroaches that infested the cupboard in his office. He was, however, somewhat put out when he found one of the cats had chosen his office chair in which to give birth. Fortunately, the secretaries promptly fell in love with the kittens, who were then all given proper homes with regular meals.

4. Currency

Matt had noted the importation of American ideas, and was not surprised when Australia decided to decimalise its currency. He was intrigued to learn that the first suggestion of the name for the major unit was 'Royal', but that this was rejected. He imagined that this might be replaced by some national name like 'Oz' or 'Roo' but no, the country settled on the well known and widely used 'Dollar'. Following bygone lessons in bookkeeping, Matt approved heartily of decimalisation, since addition in dollars and cents is far easier than that in pounds, shillings, pence, halfpence and farthings. After all, he would say, think of the difference in hassle when adding up sums like the following:

£3.14s.5¾d.	£3.72
£2.6s.3½d.	£2.31
14s.7d.	£0.73
£1.15s.11½d	£1.80

Even though one of Matt's teachers had been capable of adding up pounds, shillings and pence progressively, most people had to handle them separately and, despite the convenience of the figure twelve being divisible exactly by four and by three, Matt felt that these advantages were greatly outweighed by those of decimalisation, which also enabled the simpler handling of percentages for things like taxes and sale reductions. However, he regretted the fact that the new system centred on the ten shilling unit rather than the pound. This immediately gave the impression of doubling the price of everything, especially of expensive items like houses and cars, and he recalled his months in Europe where prices in small units of francs, etc had led to figures so extensive as to appear almost meaningless. But there, the country must make its own decision, and Matt respected it accordingly.

*****

The good news came through that the college had been granted its autonomy, and was now a university in its own right, permitted to grant its own degrees. This called for a celebration, which Matt and Joyce attended with pleasure. The following day, the Principal, shortly to be upgraded to Vice Chancellor, called Matt into the boardroom and advised him that the Board had met and decided on the need for a permanent Head of Department of Mechanical Engineering, at full professorial rank. As a matter of protocol, this vacancy would have to be advertised, but was Matt interested in applying? Matt accepted with enthusiasm, and was eventually appointed to the post. His reasoning was that there was just the possibility of the family in the UK joining them but if not, he would try to stay long enough to get the new university department off the ground and well established.

Some months after this event, a bombshell hit Matt in the shape of a telephone call from England saying that Dad had met with an accident. Apparently he was up a ladder painting the gutter when he slipped and fell, cracking his skull on the concrete path. Matt raced home, and talked over the situation with Joyce. He was torn between flying back to the UK to help, or staying to handle his family situation and the continuing development of the university. He decided on the latter because, although reasonably fit, Joyce was undoubtedly delicate, and he was not prepared to put too much stress on her.

The situation was settled a few days later when a second phone call was received to say that Dad had died in hospital. Matt felt as angry as he did sad. If only this had happened a few months later, he would have had a chance to talk to Dad of the many things he wanted to say about their life abroad. He cursed himself for having left England in the first place, and Joyce did her best to comfort him by pointing out that Dad had wished him to follow his star, and had been proud of him for so doing.

Inevitably, memories of life with Dad flashed back in Matt's mind. How he and Mike used to go to meet their father on his way home from work, and fight each other for the coveted position on his saddle as he wheeled them back on his bike. How they used to go blackberrying at Keston, and Dad, as a countryman, never bothering about snails and spiders crawling over his large hands in the process, whereas the two boys were a bit reluctant, and Val was positively unco-operative. Dear Dad, he had been such a tower of strength for them all, and so much respected by his work mates and neighbours. He had loved singing, and always made a point of attending a male voice choir, even through the war years. Matt realised, as we all do, that we have to lose our parents eventually as part of life's pattern, but it still hurts when it happens.

Matt decided that they must now return to the UK and by chance had noticed a vacancy for some educational research in London, so he applied immediately, and was accepted. Dennis and Val made a special visit, but the pleasure of meeting was overshadowed by the sad events. Dennis did offer to handle their departure arrangements if Matt wanted to fly back beforehand but, although Matt was grateful, he felt it too much to ask of someone who lived several hundred miles away. Above all, he felt that it would be far too much to ask Joyce to handle the selling of the house, packing their goods and organising the return voyage.

The university staff took Matt and Joyce out to a Chinese restaurant for a farewell meal and, much to Matt's pleasure, presented him with a boomerang on which they all signed their names. In case this should be lost by throwing it wrongly, Matt made a point of buying another boomerang to practise with.

Preparations for the return journey now went ahead without disruption. The house was sold to a new member of the growing university staff, and the house agent found the Greggs some rooms for a temporary letting. On selling his car, Matt was able to order a similar model to be collected on arrival in England, so they would be mobile from day one. This should be waiting for them at the dockside in Southampton. Dennis and Val drove them to Circular Quay in Sydney, and there they joined the sleek looking *Canberra* waiting at the quayside.

"I see they've put the funnels at the stern instead of the middle," said Steve. "I wonder why that is?"
"Don't quite know," replied Matt. "Maybe it's to keep the smoke off the swimmers in the pool."

This may have been an effect, but *Canberra* was a significant step forward in the design of ocean liners, as Matt and his family were to discover over the next three weeks. They were able to spot Dennis and Val in the crowd on the quayside, and it was clear that Val was distinctly upset and tearful, which wasn't all that surprising under the circumstances, but they had all promised to send each other airletters whenever they could.

The tugs nursed *Canberra* away to just downstream of the Harbour Bridge and she moved forward to start her home voyage, threading her way through the magnificent harbour, which was highlighted in the strong December sunshine, and out into the open sea. The slight vibration from the engines, and the reactions from the waves confirmed that *Canberra* was alive, and in her element.

This is the life, thought Matt, as he took the family to the restaurant below for lunch.

Chapter 16 – Survey (1966 – 69)

The first few days of the voyage were occupied be exploring the ship, and sampling as many as possible of the delights on offer – restaurants, bars, swimming pool, cinema, and so on – but the item that intrigued them most was the clever design of cabins to give all of them a sight of the sea. This was achieved by extending the length of the internal cabins to incorporate a window to line up with a large window at the outer end of an athwart ship's corridor. Matt loved spending hours watching the sea, and noting the range of colours it showed in different conditions of light as he tried to figure out why the wake of the ship curved outwards rather than remaining as a straight cone.

Steve pointed out the sea birds and large mammals that chose to accompany the ship on occasions, and sometimes the flying fish. True to its name, the Red Sea fluoresced at night when the passage of the ship disturbed microscopic creatures in its path. Christmas Day arrived, and Matt asked Joyce to film him swimming in the pool because he would like to assure everyone that he would never do that again on such a day – not in a European climate anyhow!

The portstops again stretched their experience into a wider world and, at Bombay, the family noted the arrival of a lean-looking dog who settled himself on the quay exactly opposite the galley. After several hours his patience was rewarded and he gratefully retrieved a large bone thrown out for him and disappeared into the darkness to devour it. A pitiful figure of a man with badly deformed arms and legs then appeared, progressing along the dock in crab-like fashion begging for alms. Matt was shocked to hear from another passenger that the parents of some families out East deliberately deformed their children's limbs in order to enhance their appeal as beggars, and looked on this as an act of kindness!

Once again, the Suez Canal provided an interval of completely calm progress, and they could not miss a visit to the pyramids. Whereas the ladies and small children were taken to the site by horse-drawn carriages, all the menfolk were invited to mount horses, or camels, for the short journey. For the first time in his life, Matt found himself on horseback, and would have been much more comfortable if the guides had let him take the reins, to give him some stability, but they insisted on leading the horse themselves. Matt was amused to note that, although they had been assured that the cost of their ticket for the visit included a tip for the guides, they all clustered round Steve and pestered him with the repeated question, "Which is my father?" obviously on the lookout for a bonus, which Matt was not unhappy to supply since he was enjoying it all so much. At the pyramids, Matt searched for evidence of the bitumen used for joining the massive stones together, but this seemed to have disappeared over time. The family, on the other hand, went inside a little way and were intrigued to find it illuminated with a fluorescent tube light – it just didn't seem quite appropriate somehow.

The Mediterranean proved surprisingly choppy for a relatively protected sea. With nightfall, by training the ship's searchlights to meet at a point just above funnel height, it was possible to see the power cable strung between Italy and Sicily. The faint glow above the peak of Mount Etna could also be detected in the dark. The results of the painstaking work of excavating the ruins of Pompeii and nearby towns were impressive, and some of the ancient graffiti on the walls even more so!

At Gibraltar, the ship turned through a complete circle in order to show off the Rock from numerous aspects, and it was noticeable how the wind dropped as soon as they moved into the massive lee. On reaching the open sea and negotiating the Bay of Biscay, the decision was made on the bridge that this would be a good opportunity to test the 'Man Overboard' procedure. A life belt was therefore thrown overboard, and the ship described a large circle, while slowing to permit the lowering of a lifeboat to effect retrieval. This was all very impressive and reassuring, but Matt noted that the whole operation took at least twenty minutes – a period that could prove fatal in a rough shark-infested sea. Despite his very deep respect for just about everything in the marine world (and he had become a Shoreline member of the Royal National Lifeboat Institution as soon as he became aware of it) he had absorbed the philosophy of aviation where speeds were so much higher, and periods available for action so much shorter. Since aviation had learnt so much from the marine world (e.g. rudders, variable-pitch propellers, etc) he thought it a good opportunity to give something back, so Matt went below and sketched out what he called a 'Crash Boat'. This was to be an enclosed lifeboat of particularly rugged construction that would be located at the stern of the vessel and, when required, occupied immediately by two duty crewmen who would lock the doors and press a button to rocket propel the boat into the wake. Once settled, the boat could instantly locate and rescue the victim within a couple of minutes of sounding the alarm. Matt vowed to forward this idea to the editor of some major newspaper as soon as he reached home. (He was as good as his word but, sadly, no response was forthcoming.)

This episode also reminded Matt of a tragic event that had occurred years back when a submarine had got into difficulties and could be seen with its stern sticking out of the sea. Valiant attempts were made to salvage it by attaching cables to the stern, but she slipped out of them and was lost with all her crew. Matt's training to foresee accidents and build in safety beforehand (for instance, the shape of an axe painted on the side of an aircraft fuselage so that rescuers could be guided to the location of a safety axe to cut open the cockpit cover and retrieve the pilot) led him to wonder whether strong points could be incorporated into the submarine's structure to accept a hook and chain lowered from a rescue ship to hold the submarine, at least until the heavy rescue vessels arrived. His thoughts also ranged over the possibilities of a number of access points to which safety hoses might be attached by divers who could then trigger valves to permit the ingress of fresh air and the exhaust of used air by means of co-axial tubes. Even if this were not feasible at great depths, it might perhaps be practicable just below the surface, as with the stricken submarine in question. However, this was not his area of specialisation, and he was sure that the naval architects would have considered it.

The landfall party on board the night before reaching Southampton was utterly memorable for the whole family (but presumably not so for some of the passengers who hit the sauce rather heavily), and the word 'Landfall' was to remain in Matt's mind. There was something about the motherhood of

England that was not always apparent until you moved away for a while. Despite the price they had had to pay in losing such a loved member of the family in their absence, it was an experience that had matured them all.

Stepping ashore the next morning promoted strangely restrained emotions, but they were able to go through the process of locating their baggage without too much hassle. What was annoying was that their promised new car was just not there! However, a driver from the motor company did make contact with them, and was able to cram them all into his station wagon and transport them to the main office in London. His non-stop supply of jokes, updating them on recent happenings in the old country, helped to calm their irritation, but then Matt had to load his new car and negotiate his way out of London to Brickhill based on five-year old memories, whereas he had memorised his route from Southampton. Still, this was nowhere near enough to spoil the magic of the occasion, and they were all greeted with deep affection and relief by Mum and Dorothy when they reached home.

Accommodation was a bit tight all round, but relieved when a corridor was transformed into a bedroom for the children. Toby had long since joined his ancestors in that great cattery in the sky, but the feeling was that they were not quite ready to take on board a replacement pet.

Once they had settled in to their new home routine, Matt took the train to London and reported at the office of the survey company. It transpired that their staff comprised mainly ex-teachers and arts graduates, but few with any strong scientific background, so they were particularly pleased to welcome Matt on board. The plan of action they devised between them was for Matt to:

a) list all the UK universities and colleges that incorporated departments covering the various branches of engineering and applied science, then visit as many of these as practicable and prepare a comprehensive report, and,
b) where possible, correlate any relationships between academic performance and subsequent professional success.

They all recognised the extent of work that could be involved in either of these projects, let alone both, but it was agreed that the results of a three-year study would indicate where and how to proceed further. To Matt, the beauty of this task, apart from the intrinsic interest he had in technical education, was the fact that he could work from home and touch base with the London office only when necessary, probably once a month. Although he did not feel comfortable in dragging Joyce around the country in view of her delicate health, he did envisage taking her on some of his London trips for a meal out, and perhaps a show afterwards.

They were both pleased that Steve had gained entry to a college to read natural history, and also to start studying photography between terms. Avril had set her heart on ballet dancing, but when she found that she was not built for it, decided instead to study optometry rather than go in for general nursing as her mother had done. Both Steve and Avril had moved out to student lodgings, so much more space had become available in *High Firs*.

One black day, came the tragic news of the terrible loss of teachers and children when a slag heap moved and engulfed the school at Aberfan. As Matt read this in the paper on the London train, he reflected that it is painful enough to lose parents, as he had already experienced, and that it must be much worse to lose a beloved partner, but to lose children who have only just started in life must be heartbreaking in the extreme. He ached for the bereaved parents in Wales, and only wished there was something he could do.

Spirits rose a little with the appearance of the Harrier jump jet at Farnborough. However, the losses of the three US astronauts, and later of Donald Campbell on Lake Coniston the following year, were extra sad blows to anyone interested in technological challenges, particularly, as in Matt's case, where spaceflight and water transport were involved.

Steve's week-long visit home, complete with a mountain of unwashed laundry, coincided with the announcement that 22,000 acres of north Buckinghamshire were designated as the site of a new city to take its name, Milton Keynes, from one of the villages it would incorporate. The Greggs viewed this with some concern. The concept was exciting enough, but it would be perilously close to Brickhill, and there was no knowing what overall effects it might have. Only time would tell.

A moment arrived that Matt and Joyce had been looking forward to for months, when they were able to take up the invitation to visit Brenda and Norman in Norwich. When Matt recounted their drive from Bedfordshire to East Anglia, and how they had had to stop at Wymondham to spend a penny, he was greeted with gales of laughter. Puzzled, he enquired what he had said wrong. "Oh," was the reply, "we pronounce 'Wymondham' as 'Windham'." "And also," they added, "'Costessy' as 'Cossey'." "Well, thanks for telling me," replied Matt. "I won't make that mistake again." Secretly, he was pleased that he had noticed that the correct spelling of Hethersett did not contain the letter 'a' as some of his friends had concluded when describing the fuel pipe-line there. We live and we learn.

Norman and Brenda took them on a brief tour of the city, after explaining that the first time the name of the city was recorded in writing was in the form of 'NORWIC' on a coin minted around the year 930. The most obvious landmarks, of course, are the cathedral spire, 315 ft above ground according to Norman, and the square bulk of the castle. A historical fact that intrigued Matt was that the cathedral, which was built of Norfolk flints, was then coated with white stone brought over from Caen, which he had passed through during the invasion. In order to complete the final leg of the stone's journey, a small canal was built from the River Wensum to the cathedral itself, with a now-famous gateway built in the 15th Century to guard the canal. This point was used later for crossing the river by Pull, the ferryman. Walking within the cathedral close, they were shown the grave of Nurse Edith Cavell, executed by the Germans during the war, in 1916, and also the noted King Edward VI School, together with a statue of its famous pupil, Lord Nelson. It was striking that this statue stood back to back with that of the Duke of Wellington, but Norman explained that the two celebrities, who had met only once, had not been on friendly terms with each other! Matt and Joyce were shown the wealth of architectural treasures within the cathedral itself, including the colourful stained-glass west window, and told about the grave of Thomas Gooding of Tudor times who was buried vertically so that, on resurrection day, he had merely to step forward, rather than stand up!

The thoroughfare that seemed to typify Norwich and its history was the cobble-paved Elm Hill leading from the ancient elm tree down to Wensum

Street, comprising antique shops and art galleries. Norman pointed out the Guildhall at the edge of the market dating back to the 15th Century. He indicated that, some years back, its committee had debated whether or not it should be pulled down, the voting being exactly 50/50 for and against. However, the fact that it was still there was because of the chairman's casting vote, and it is still the home of the Spanish Admiral's sword presented by Nelson.

The nearby City Hall was opened by King George VI in 1938, the intention being that two wings of building would be added. The southern wing was completed to house the fire and police stations, but the northern wing was abandoned with the outbreak of war. Norman pointed out the rolled steel joists still protruding from the wall ready for the intended wing. There was insufficient time to visit the Mustard and Bridewell Museums, or Strangers Hall, but Norman recommended them as worth seeing later. He also mentioned a Doctor Rigby who devised a method of removing gallstones – something of a problem in Norfolk in the days before anaesthetics – so rapidly that the patient hardly had time to draw breath to scream before the job was done! Interestingly, Dr Rigby's errand boy, John Crome, who was the subject of a practical joke involving a skeleton being placed in his bed by some medical students, promptly responded by dropping the same skeleton onto the heads of the said students. Dismissal followed, with Crome becoming apprenticed to a sign writer, and eventually forming the Norwich School of Painting. Norman also made brief reference to the rebellion led by Robert Kett back in 1549, and advised them to check on their way back on the ancient oak tree still in existence at Wymondham where Kett gathered his force of 20,000 yeoman farmers to fight the enclosures of common land. One recent addition giving rise to civic pride was Anglia Square, with its complex of shops, and the concrete-and-glass headquarters of the Stationery Office.

Norman explained that the reason why the Norwich football team was nicknamed 'The Canaries' was because their strip was similar to the colourful pet birds that the weavers brought with them when migrating from the Netherlands back in the 16th Century. Some old houses with extra large windows to let in the light for weaving were still in existence in the city.

Joyce and Brenda enjoyed a wonderful chat over old, and recent, times, and both were interested to hear of Professor Barnard's success with heart transplants in South Africa. They had both seen so many people succumb to heart disease when a transplant, had it been possible then, might have given them several more years of good quality life. The next advances, they hoped, would be in the treatment of tuberculosis and, as always, cancer.

Before leaving, they were delighted to meet up with Dulcie who was on a flying visit from her nursing course in London. Geoffrey, now sixteen, was specialising as much as he could in the science subjects at school.

Back home, some very welcome news arrived from Australia to the effect that Dennis had been offered a post in an Oxford college to lecture on the history of the Pacific region, and that he and Val expected to arrive within a few weeks. Harry was completing a course in economics, and expected to join them after graduation, at least for a year until he found a permanent position somewhere. Arrangements were made for temporary accommodation at *High Firs*, then Matt and Joyce drove to Heathrow to meet the flight from Sydney. After quickly settling in their guests, they all repaired to the local *Barge Inn* and compared notes on events since their last meeting at Circular Quay. Matt and Dennis were particularly engrossed with the forthcoming maiden flight of the magnificent supersonic Concorde, discussing it with such enthusiasm that Joyce and Val stopped their chatting for a moment, noted the animated conversation at the bar and remarked, "And they say *women* can talk!"

* * * * *

Matt found, somewhat to his surprise, that his hours of checking the work of students had enabled him to scan a written page and to note spelling errors instantly, almost as if they jumped out at him. This was not something he had striven to achieve – it was just the result of experience. The funny thing was that he could not do this with his own work. No doubt this was because he viewed his writing as the way he had intended to write it rather than the way he actually had.

As a boy he had evolved the idea that everybody in the world should be encouraged to learn two languages, one their own mother tongue, and the other an international language so that everyone could converse with everyone else. True, there had been attempts to devise such a common language – Esperanto, for example – but it just did not succeed. He noted that aviation, with its pressing need for common understanding of flight control and its widening contacts worldwide, could no longer wait for such an ideal situation to reach fruition, and so had arbitrarily accepted the most practicable language in existence. And that proved to be English, which has the greatest number of words of any language, giving innumerable shades of meaning. Admittedly, it had its drawbacks (cough, dough, rough, through, etc, spring to mind) but Matt considered that this was the optimal choice. On next meeting his colleague John, Matt asked him, as an Arts man, on his views regarding this idea. John agreed readily, but pointed out that Spanish was also used in many countries around the world, and might be perceived by some as a potential international language.

From his limited knowledge of French, Matt recognised its beautiful romantic nature and the aptness of certain words which could not be matched in English (so we adopted them!), words and phrases like 'au fait', 'bon mot', 'chic', 'de trop', 'déjà vu', 'raison d'être', 'rendezvous', 'tête à tête', and 'touché', and in aviation, 'aileron', 'échelon', 'fuselage' and 'nacelle'. What a pity we hadn't also taken on board the delightful 'le crépuscule' for dusk, 'la dessinatrice' for draughtswoman, and 'la finesse' for fineness ratio. One typical British prejudice is that many French terms are far more lengthy than their English counterparts, as in: anti-clockwise – dans le sens inverse des aiguilles d'une montre, and nosewheel steering – le système de guidage de la roulette de nez. On the other hand, the reverse holds true in some instances, as in: centre of gravity position – le centrage, and land on a carrier deck – apponter.

However, Matt felt that the cedilla and various accents were just too fussy to be practical, and the insistence of applying gender to every article meant an inexcusable memory task. Much better to learn a few rules than committing a host of such detail to memory. Why on Earth, for example, should a pencil be classed as masculine, and a pen as feminine? And why were the French so averse to introducing new words into their vocabulary? In

English, we did it all the time, and that helped to give the language such richness. Surely any language that insulates itself from other tongues can hardly expect to be adopted internationally.

Matt knew almost nothing about German, but it seemed such a heavy language, with great long composite nouns that it didn't seem to attract. Nevertheless, here again many terms had entered our own vocabulary like 'angst', 'auf wiedersehen', 'blitz', 'ersatz', 'gesundheit', 'kaput', 'lederhosen', 'lieder', 'luftwaffe' and 'panzer'. So Matt congratulated himself on knowing both his mother tongue and an international language (certainly as far as aviation is concerned) at one and the same time.

By this time there were certain words and phrases that Matt had come to treasure, and also certain errors that irritated him intensely. For example, he felt that we ought to stress the fact that some words are singular, and others plural, and that they should not be mixed up. He was mortified to note that, even in some of the better newspapers where the editors were normally masterly in handling the language, one could sometimes find such errors. So he drafted out the following list which he thought should be pinned up in all schools (and some editorial offices!):

Singular	Plural
phenomenon	phenomena
criterion	criteria
medium	media
stratum	strata
graffito	graffiti

He therefore struck out written phrases such as 'this criteria' and 'this phenomena', with a thick red pen whenever he came across them. He did not include 'museum' and 'musea', nor 'referendum' and 'referenda' but wondered if they should have been, until he referred to the dictionary. Also he drew attention to 'different from' and not 'different to' (although in America it was 'different than'!). Furthermore he added that 'none' stood for no one, or not one, and that as singular it should be followed by 'is' and not 'are'.

212

He was also put out by the apparent lack of teaching of the use of the apostrophe. You only had to visit a street market to find examples of "Lovely sweet apple's". Surely it would take but a few minutes to have had the perpetrators of such errors taught at school that the apostrophe indicated possession, and was not needed for plural, giving such correct examples as "the apple's skin" (or "the apples' skin" if there were more than one of them) and not to use an apostrophe when writing about apples in plural rather than something they possess. The only major exception to this rule that Matt had noted was concerned with the word 'it', where the apostrophe is used for the contraction of 'it is' or 'it has' (i.e. it's), but is not used to indicate possession. He remembered this easily from the phrase 'It's time the dog had its bath'.

But then, Matt had the impression that there had been a period in schooling when the accent had been on the touchy/feely shapes of sentences rather than the nitty gritty of grammar and spelling. Matt even wondered if the teachers themselves were sufficiently aware of these rules! Thankfully, things seemed to be swinging back to a more logical approach, although there were still some diehards prepared to ignore spelling in homework or exam papers instead of guiding the pupils at every possible opportunity. Matt always counselled students to have a dictionary readily available (at their elbow, not on a shelf the other side of the room) and to consult it every time they had the slightest doubt about the spelling or meaning of a word. With the coming of the word processor, of course, this was greatly helped by the inbuilt spell-checker.

Matt had also reached the conclusion that there were four stages of fluency:

1. You don't talk much because you have so little to say compared with those around you.
2. You talk a lot because you are reasonably well versed in one aspect of the topic of conversation, but not smart enough to know when to keep quiet and listen to others.
3. You don't talk much because you can see both sides of the argument at the same time.
4. You are able to talk at length because you understand in depth the various aspects of a multitude of issues.

Matt liked to think that he was graduating steadily from stage 3 towards stage 4.

This comfortable life was jolted by a telephone call from Joyce telling him that Mum had been taken ill and was now in hospital. The cause, apparently, was a heart attack, but she rallied and looked set for recovery. However, a second attack followed, and Mum lost her battle with life. Even if Dad had still been around, it seemed unlikely that she could have survived, but without him, she had little urge to continue living, despite her love for her family. Her time had come, and she knew it, accepting it gracefully.

With the moon walk of Neil Armstrong, and his successful return with his colleagues to Earth, the family were glued to the TV set for hours on end, going without meals until Joyce public-spiritedly volunteered to organise them. All knew that this was an historic moment in the development of mankind, and the beginnings of unimaginable explorations in the future. Matt had always admired the way in which scientists had been able to build up quite a detailed picture of the structure of the universe and the relative movements of the bodies within it, and always from one point only, our own Earth. But now, there would be cameras, sensitive instruments and even human crews travelling from the Earth to give us a much deeper insight into the whole magnificent immensity of space. It was thrilling, exciting and slightly awesome, but something we just had to do. There was no going back. If the Earth as we know it is eventually doomed for destruction in several million years to come, then the human race must either accept annihilation or search for alternative planets to colonise, and hopefully not wreck in the process! "One small step for man," as Neil Armstrong put it, it may well have been, but the potential for the future was mind-boggling. Matt felt so humble and grateful to have been living at this momentous occasion, and he was certain that both Dad and Mike would have been equally thrilled.

The long-awaited maiden flight of Concorde also gave the Gregg family cause for pride, whereas the discovery of oil and gas in the North Sea gave an even greater impetus to the confidence of the nation, as Matt noted with considerable satisfaction.

Matt was relieved to be able to complete his educational survey, and to submit a report on it to his parent company. It was received well, and would undoubtedly be valuable to himself as a background picture of the educational system in engineering within the UK.

Although Matt had made a number of useful contacts during his touring of the engineering departments of the UK universities and colleges, the completion of his contract with the educational survey company coincided with a trough in the vacancies in the academic world due to financial constraints, so Matt's first priority was to find a temporary job. While following up possible leads, he found that he needed to attend the local employment exchange for some financial support. This proved quite distasteful to him. He was on the dole, and didn't like having to take handouts. When asked at the desk what his previous job had been, he quite innocently replied that he had been a Professor of Engineering. He hadn't wanted to sound at all big-headed – it was just the truth. "Right," said the clerk, "You line up over there!" Matt struck up friendships with others in the queue, and they all shared the same misery at seemingly not being worthy of employment. A couple of interviews did come his way from some remote colleges, but, to Matt's chagrin, the interviewers seemed to attach no importance whatever to his senior academic experience overseas – it seemed that only work in the UK itself was significant. This was just the sort of ignorance about the Commonwealth that Matt himself had sought to rectify by going to Australia in the first place.

Some people took their unemployment trauma quite badly, and suffered depression, but Matt was determined to use the time positively. He therefore set about completing his notes on spaceflight propulsion and combustion calculations, since he would probably never have an opportunity like this again to concentrate on these projects in such a detailed manner.

Just when job opportunities seemed to be non-existent, a telephone call came from a friend in the university in which Matt had previously worked, to the effect that a vacancy had arisen there for a senior lecturer, and would Matt be prepared to consider it while he waited for something a little better? Would he not? The rank didn't matter to him, and in any case he didn't fancy more administrative duties, much as he had enjoyed them in Australia.

What he wanted now was as much time as possible to spend on research, lecturing and writing. The interview proved enjoyable because the board members were all known to him, and they in turn were fully aware of Matt's background and achievements. They did make the point that they would look to him to continue publication, and Matt happily agreed to embark on a series of textbooks covering thermodynamics, space propulsion and, of course, his favourite subject of fuel technology. Lubricants were to be covered by another member of staff who lectured on tribology. Although there had been quite a few changes of staff in the interim, Matt was able to meet up with many old friends, and fulfil his promise that he would one day return to visit them.

The relief and pleasure experienced by the Gregg family was topped by a visit from Steve, complete with Margaret, an attractive librarian assistant he'd met when spending part of his summer work experience at a natural history museum.

Chapter 17 – Return (1969 – 79)

Now that they had settled in long-term at *High Firs*, Matt and Joyce set about refurbishing the place and improving the facilities. Matt first redesigned the kitchen, guided closely by Joyce, and then tackled the stripping out of the old, and fitting of the new. The bedroom was redecorated and recarpeted, as was Dorothy's room, in which they also installed a hand basin. They eventually coped with the expense of extensions in the form of a laundry and a porch, and the renovation of the attractive summerhouse in the back garden. Routine maintenance also had to be continued with the front cottage, which was then occupied by some students from the university.

The happy news arrived from Brenda that daughter Dulcie was to be married to Robert, a recent medical graduate. The Greggs were invited to the wedding, so Matt and Joyce were able to enjoy another memorable weekend in Norwich.

Mindful of his university commitment to embark on a series of textbooks, Matt completed his useful little handbook outlining the calculations required to determine the mixture strengths and exhaust products involved in combustion, plus the resulting energies, temperatures and efficiencies for a variety of fuels of known composition. He included many examples to show how the information could be used easily. The Combustion Calculations booklet was then produced cheaply in photocopied form, and sold out rapidly. Relative to the amount of work involved in producing textbooks, of course, the financial gain is minimal, but it is a good discipline for the writer, and can be of direct use to the reader.

With increased free time from preparing and giving lectures, Matt was now able to contemplate some further research, and he chose two projects. The first was to collate his notes and complete his book on Spaceflight Propulsion, for which he managed to find a willing publisher, and the second was the continuation of the ignition work that he had started some years earlier, bearing in mind the fuels alternative to the conventional fossil-based gas, oil and coal products in view of the shortages that must inevitably occur when these resources are spent.

Matt had gained much experience with the standard method of measuring spontaneous ignition temperature of fuels, in which a droplet of fuel is inserted into the hot air within a flask, and the lowest temperature found at which the fuel would ignite. These results were interesting in giving a comparison between fuels, including the alternatives. However, what concerned Matt was that the test was conducted with an open flask, i.e. at atmospheric pressure, and the delay before ignition occurred could be several seconds. In an engine that is spark knocking, on the other hand, the pressures are, of course, quite high, and only a few milliseconds are available for ignition to occur. So Matt devised a rig in which a droplet of fuel falls under gravity through a heated furnace, with optical detection of when the droplet enters the furnace and when it ignites, thus giving the delay in milliseconds. Also, the whole rig was encased in a sealed box, so permitting the high levels of pressure required.

Matt had always been aware of the finite nature of fossil fuels, so he also surveyed all the likely alternatives that he could think of, including alcohols, ethers, vegetable oils, ammonia, liquefied gases, coal-derived liquids, and hydrogen itself. This appeared to him to be such an important issue for the future that he started to prepare a textbook on that subject also. 'This ought to keep the university happy,' he surmised. Whereas the earlier targets had been for maximum power, the picture was now changing for maximum economy. At all events, he was content that a whole raft of interesting and useful research stretched out in front of them all, and that he could count on supporting a number of students working in this field for their higher degrees.

Matt continued to monitor events in the aerospace world, and became well aware of the concern that gripped the nation for four whole days when the news broke of the struggle by the crew of Apollo 13 to cope with bringing their damaged craft safely back from Moon to Earth. There were general fears that this would be a follow-up to the fatal accident to the American astronauts in 1967 when fire broke out on the launch pad. Thankfully, all went well this time. Matt was gratified to see further confidence in aerospace restored with the first landing of a Boeing Jumbo Jet at Heathrow, and later the successful buggy ride on the Moon by the astronauts of Apollo 15.

More good news arrived from Norwich to the effect that Dulcie had been delivered successfully of a daughter, to be named Wendy, and that the presence of Matt and Joyce was requested at the Christening, so another pleasant stay was enjoyed with Brenda and her family.

Having witnessed the introduction of decimal currency in Australia, Matt was delighted to see its adoption in the UK, and also with the one pound, rather than the ten shillings, as its major unit. He was glad about the earlier demise of the farthing, and hoped that the halfpenny would soon follow suit (which in fact it did). The term 'new pence' did not last long either, but what puzzled Matt was the use of the term 'one pence' instead of the familiar 'one penny'.

However, the main item of interest in the family now was the impending marriage of Steve to his attractive girlfriend, Margaret. This time, of course, there were none of the wartime restrictions that had constrained the accompanying festivities of the marriages of Matt and Joyce, or of Dennis and Val, so hearts were correspondingly lighter, and the pleasure gained by all involved that much greater. Steve's interest in natural history and the countryside seemed to give him a solid confidence, and he came through the ceremony and the reception showing few signs of nervousness. For her part, Margaret looked an absolute picture and, although it is traditional for the bride's mother to shed a tear or two on such an occasion, Matt found it difficult to disperse a lump in his own throat, as the sight of the radiant Margaret reminded him so strongly of the appearance of Joyce when they married back in 1943. It was also a great pleasure to welcome Brenda and Norman to the ceremony. Steve had organised one of his former fellow students to act as best man and, after some witty speeches, the happy couple departed on their honeymoon to the south of France.

The family had only just waved goodbye when the depressing news came through of the crash of the Trident aircraft at Heathrow, with the loss of all on board. It was at times like this that Matt had to remind himself that, although aviation tragedies had occurred with sickening frequency, they had to be seen in context of the thousands of successful flights taking place around the world every day, and that, in relation to all other means of transport, aviation enjoyed a remarkably high safety record. He just hoped that the general public would bear this in mind when confronted with dramatic news announcements of such disasters.

"Don't suppose we've heard from Steve and Margaret yet," remarked Matt one lunchtime, later changing his line of thinking. "They won't be in a mood to write postcards at this early stage. In fact," he added wickedly, "they might even not be up yet! Do you remember our first days on honeymoon, love?"
"Remember them? – I'll never forget them! I was a changed woman then, as you well know."

They smiled at each other in fond remembrance, and Matt stood up to stand behind his wife, kiss her head, and slip his hand down the front of her dress to fondle that special breast. "You're wicked, that's what," Joyce murmured as she lifted her head to kiss him in return, smiling in the recognition of their shared love.

Steve and Margaret duly returned looking very pleased with themselves, and with each other, and since Steve had secured a job in a local museum, they moved into the cottage following the departure of the student tenants. Furthermore, the need to make fairly frequent sorties to zoos and arboretums for securing specimens and taking photographs meant that individual transport was necessary. So Steve hunted out an old station wagon and, with Matt's help, brought it up to roadworthy condition.

Whether or not Steve's obviously successful wedding acted as a catalyst or not is unknown, but Avril indicated that she had narrowed her retinue of boyfriends down to one, and that he, Mark, a slightly shy anaesthetist, had proposed to her after several appointments (not all of them perhaps strictly necessary) at the optometrists.

Matt felt vindicated when the 1973 fuel crisis brought home to the nation the vulnerability of fossil-based fuels, so much so that he devised a series of rig and engine research projects to check the practical suitability of the different alternative fuels that he had been studying. He was already collecting much information on the spontaneous-ignition characteristics of these fuels, which would be of vital importance in assessing their resistance to spark knock and diesel knock. He also resolved to complete his book on the subject as soon as he had collected enough material to present a balanced picture.

"Why is it," asked Joyce over the breakfast table, "that we have to go short of fuel? And why do we all have to put up with working only three days a week?"

"It's a bit complicated, love, but the countries that have the oil have decided to handle the production themselves, and the price has shot up. Also the miners have gone on strike, and there's a general crisis in the economy, so it all adds up to a major problem."

"Do you think that the miners should have gone on strike?"

"Difficult to say without knowing all the facts, but I think that any strike is a pity. Trade Unions are essential, of course, to protect the workers against any rogue managers. But on the other side of the coin, somebody has to act as manager, and be allowed to manage without deliberate attempts to sabotage the company. I do think a strike is sad, because it means loss of income to all the company staff, and loss of products to the customer. I don't really understand why management and shop floor are so suspicious of each other. If they really want conflict, why don't they act together as a team and challenge their rival companies? I'm also unhappy about this picketing business – it smacks to me of intimidation. I only wish they would either scrap it or insist that someone from management is also present to put over their viewpoints as well. But, of course, there's not a cat in hell's chance of that happening."

"I see what you mean," replied Joyce. "If I wanted to go to work I would hate to have to meet intimidation from people who didn't happen to agree with me. Surely we all deserve the right to choose?"

"Course we do. It all has to end with discussion round a table, so why the blazes can't that happen sooner rather than later? But there again, we probably don't know all the facts. Pity though."

Some weeks later came the news of the crash of the Concordski, the Russian equivalent of Concorde, another tragic step in the development of aviation. However, the family was then cheered to hear that Avril and Mark had also decided to marry, so they once more went into their wedding mode with all the usual arrangements. Although the Greggs were by now well versed in, and equipped for, the wedding scene, Joyce considered it would be highly appropriate to acquire a completely new outfit for the occasion, including, of course, matching shoes and handbag, together with some confection of a picture hat. Matt allowed himself to agree to all this fuss and bother, and was even persuaded to invest in a new suit, shirt and tie. He may not have emerged quite as the Beau Brummel of Brickhill, but he was undoubtedly suitably dressed to perform as the father of the bride.

Once again, Matt reckoned he could generate as much tear fluid as Joyce, the bride's mother, and they were both quite dewy-eyed as they waved the newlyweds off on their Italian honeymoon. This happy event also gave another opportunity to welcome Brenda and Norman to Brickhill.

On their return, the newlyweds were to move into a flat in the village of Histon, as Mark had been accepted for a post in the main hospital in Cambridge. It was at this point that Matt and Joyce decided to transfer the cottage to Steve and Avril, to share any future rental income equally. The only caveat was that they would now be responsible for the maintenance and insurance of the property but, living next door, Matt would be able to give a helping hand when required.

The Gregg family was then delighted to learn that Margaret was now expecting, and everyone looked forward to a new arrival. Although Margaret retained her health well during her pregnancy, she was grateful to both Joyce and Aunt Dot for helping out with the more strenuous household chores, and so was Steve. He was experienced in the routine of birth and nurture in the animal world, but it's a bit different when your own flesh and blood is directly involved. So it was a relief as well as pleasure to him, and to the whole family, when Margaret was safely delivered of a healthy son, who was promptly named James Michael. An extra thrill for Matt and Joyce, of course, was that they were now, almost unbelievably, Grandparents! This situation seemed almost to have crept up on them when they weren't looking!

Once again, a happy occasion in Matt's life was followed by a world-shaking event, this time with the crash of a DC10 aircraft outside Paris, claimed to be the worst disaster in aviation history, with the loss of all 344 passengers. This situation was not eased the following year when a tube train hit the buffers at Moorgate underground station, killing 35 people. The inevitable enquiry followed, with recommendations for further safety measures, but Matt wondered why these could not have been considered beforehand.

It was a pleasure for the Greggs when a letter arrived from Reg, after all these years:

> Dear Matt and Joyce,
>
> I thought you would like to know that, after retiring from the Air Force, I moved with my family to America, and that I've been able to find a plum job in the local aircraft works.
>
> The pay is excellent, and we have been able to afford a very nice house, with plenty of garden and even a swimming pool! Plus a car big enough to play tennis in! In fact, life is very good for us all here – but I do miss those English country pubs with all the old-fashioned agricultural bits and pieces arranged around the walls.
>
> Hope all is well with you and yours.
>
> Sincerely
>
> Reg

Matt replied to the effect that his own car was just about big enough to play tiddlywinks inside, but that he would drink a toast on Reg's behalf in one of these very country pubs he so missed.

Matt and Joyce decided on a quiet holiday in the West Country, and so motored down to the tiny resort of Mousehole (pronounced 'Mousl'), and stayed in a house overlooking the bay. One highlight of their stay was the lobster soup served on their first night – with a flavour so delicious that it became permanently locked into Matt's memory. They were intrigued to visit Lands End, where it seemed that every house had a lean-to workshop in order to turn out a continuous stream of miniature lighthouses, bowls and other curios for the holiday trade. Matt couldn't resist buying one of these lighthouses, and later wired it up at home, quickly finding that the heat of the bulb damaged the plastic window, which meant drilling ventilation holes in it.

The coastline in that part of the country is craggy, with a rugged charm of its own, but Matt and Joyce were equally enamoured with the sheer peace and tranquillity of sitting by the car on Dartmoor. One charming and thoroughly surprising feature of this part of their holiday was realised when they stumbled over the spectacular Lydford Gorge that seems to suddenly appear as an isolated ravine within the general flatness of the area. They revelled in the walk through the oak woods, and the side of the River Lyd as it plunges over the 90ft deep White Lady waterfall, and then into the Devil's Cauldron of multiple whirlpools.

The holiday brought benefit to them both, and they settled down at home with renewed vigour. But while Matt concentrated again on his Alternative Fuels book, he was delighted to learn that supplies of North Sea oil had started to arrive in Scotland, and that the Queen subsequently opened a North Sea pipe-line. This provided necessary breathing time, thought Matt, although he knew we must still keep pressing on with the alternative fuel research, because North Sea oil and gas would not last forever, and when the politicians eventually asked for advice on alternatives, we must be ready to give them well-researched answers.

It was a proud moment for the aviation fraternity in general, including Matt, to learn that Concorde had made her first commercial flights from London and Paris, although the sonic boom had become something of a problem such that supersonic flight was limited to over ocean. But Matt and his family were even prouder to learn that Avril was now expecting, and they went into their rehearsed routine of preparing to welcome the infant. There had been a slight problem of blood pressure with Avril, but it was brought under control, and baby Janet was delivered safely in the Cambridge hospital, anxiously overseen by Mark. This was the Gregg's first granddaughter, of course, and Matt looked forward to the special relationship that can develop with a granddad, much as he loved Jamey.

Matt noted that the popularity of air transport was continuing to grow steadily, but there was still the occasional tragedy, even though the overall safety level was reassuring. However this took something of a knock with the news of the collision of two Jumbo Jets in the Canary Islands, the loss of 574 lives representing an air disaster even worse than the DC10 tragedy. The oil world also suffered a blow the following year with the spillage of many thousands of tons of oil from the *Amoco Cadiz* supertanker into the English Channel, with serious repercussions on marine life. Matt concluded that things didn't seem to get any easier in the aero and oil worlds, although they were both so fascinating to work in.

He was excited at the prospects of a new approach to politics after a series of strikes when Margaret Thatcher was voted in as the first female Prime Minister in this country. He thought, 'now we'll see how things go with a woman running the country. Can't be much worse than before, surely!'

Matt and Steve would occasionally take a walk together in the local woods, where the conversation was free to range far and wide over the topics of the day. On this occasion, Matt was seeking advice on photography, and how to gain the maximum visual effect. He had an automatic camera, so most of the decisions on exposure were made for him, but he sought guidance on matters of composition and framing of the picture.

On the way back home, Steve asked whether Concorde had to use a special fuel. "Not at all," replied his father. "It flies on aviation kerosine just like any other jet aircraft. In fact, both civil and military jets use the same fuel too, except that the civils rely on heat or alcohol injection to stop any ice build up in the fuel filters, whereas the military prefer to use an additive in the fuel itself to do this job. You try to keep as much water out as possible but you can't stop it all. Incidentally, that additive also stops the growth of microbes in the mixture of fuel and water, so this will be of particular interest to you."

Steve then enquired about the fuels we would have to use for cars in the future. Matt explained that a rough idea of the so-called 'life index' of existing fossil fuels on a worldwide basis could be gauged by dividing the figure for the known resources by the current annual consumption. Assuming that both remained constant, this gave a life index of about 40 years for oil, 60 for natural gas, and 280 for coal. But, of course, consumption rates would be expected to rise steadily over the years, and more reserves would almost certainly be found. So although there was some time to search for alternatives, Matt still felt this to be urgent.

As regards the likely contenders for alternatives, Matt reckoned that the liquefied hydrocarbon gases, LNG and LPG, and also the alcohols, would be suited to the present petrol-driven cars, but that diesel fuel could be supplemented by biofuels.

"And this is where you people come in again, Steve," Matt added. "You can advise us on the best crops to use for diesel engines. I believe rapeseed oil is promising, but there may well be other plants we could use. What excites me about them is that the carbon dioxide emitted from burning one crop could be absorbed during the growth of the following crop, so the whole system is carbon neutral. It's interesting to see how things have changed over the years. When I started in this business, our main target was the power output of the engine. Then, after the fuel crisis, we turned to fuel economy and now to alternatives. I'm sure that soon we'll have to concentrate on emissions, and eventually adopt hydrogen so that the engine generates water vapour only with no carbon dioxide, but there'll be a lot of problems to sort out in handling the stuff."

Matt continued, "You know, I think this oil age of ours will have to give way to hydrogen and solar power in about fifty years or so, and since it's reckoned to have started in 1859 with Colonel Drake's well in the States, it represents a blip of only about two hundred years in all the thousands of years of human life. Mind you, there was a first oil age from about 2500 BC when the ancients used oil seepages for building materials and fuels, but practical applications fell out of favour in 1000 BC and led to stagnation – which was good news for us otherwise they might have used it all up before now."

Steve was intrigued to hear Matt's references to microbes and biofuels, because it seemed that here were professional links between their two areas of specialisation. At this rate, they would have much to discuss between them in the future.

The decade closed quite happily for Matt with the publication of his book on Alternative Fuels – probably the first of its kind. The 'first in the field' feeling was satisfying, but the sales were not outstanding. No doubt the public was still not ready to face up to the urgency of the situation yet, thought Matt. Supplies of conventional fuel had settled down again after the 1973 crisis, and the high prices had become reluctantly accepted. Matt looked on his book as an early step towards the public and political understanding that must inevitably arise.

Chapter 18 – Loss (1980 – 89)

The opening of the new decade found Matt becoming a little concerned about Joyce since she seemed to be withdrawn and depressed. After a few days of this he tackled her about it while they were washing up after dinner. "Is anything the matter, love? You seem to be a little quiet lately. That wretched TB hasn't flared up again, has it?"
"No, it's not that. To tell you the truth, dear… " and here she started to weep silently.

Matt put down the teacloth, and led her into the lounge. "Now, let's sit down here and talk it over," he said firmly but kindly. Joyce quietened down, and then explained as best she could. "To put it bluntly, I've lost my libido, as medical people call it. I love you just as much as I ever have, but I just can't look forward to lovemaking with you like I used to. I suppose it's due to that surgery catching up on me. I'm happy for you to fondle me, and I can do my best to make you happy with my hands, but that's as far as I feel I want to go. I know that lovemaking is so important in a marriage – of course I do – but I wonder if I'm any good to you now, and whether you want to go with someone else for comfort. You see… "

Matt stopped her right there. "Look, my love. I understand exactly what you are saying, but I want you to understand exactly what I say now. You are my wife, and my love. You are the one I always wanted from the day we met. And as you know, I've never been with anyone else before that day, or since, and not even after you had your surgery. To be honest, I do value sexual relief, as you know, because it certainly helps me to cope with life, and to achieve reasonable success at work, but as long as I can cuddle you when I need to, and you can comfort me in the way you say, then I'll always consider myself as a very lucky man indeed. I could go on talking like this all night, but I'll just say that I love you deeply, and I treasure you because we both know I'm the marrying kind. I need a loving partner, and you're it! Satisfied now?"

"Thank you, Matt, love. I feel much better now. I'm devoted to you, and I'll prove it as often as you want me to. That's a promise." "Right." said Matt. "Now let's finish that wretched washing-up. You wash, I'll wipe, and we can both pick up the pieces later."

* * * * *

General interest in space flight flared up again with the launch of the US Space Shuttle Columbia, giving rise to more discussion on the benefits that could accrue from a greater knowledge of space, and the ability to travel within it. Even more interest was shown in the mounting of the first London Marathon, and the nation was glued to its TV sets, resulting in a tangible air of shared healthy pleasure and generosity, with a public spirit reminiscent of wartime.

The invasion of the Falklands by Argentina created a wave of anger around the nation. However, Mrs Thatcher took control much as Winston Churchill would have done, and despatched a force to sort the matter out with little delay. The shortage of troopships was overcome by commandeering some of the UK's leisure vessels, including the *QE2* and the *Canberra*. Matt, like many others, held his breath in the hope that these two lovely ships would not be damaged or lost. Fortunately, they came through unscathed, but other ships and their fighting complement were lost on both sides.

The raising of the Tudor ship, *Mary Rose*, also had the nation glued to TV, especially when the lifting gear seemed to fail just as the hull broke the surface. After such inspiring developments, gloom descended with the news of the growth in unemployment numbers. Matt was so grateful to have secured such an interesting job – in fact, to have a job at all, rather than queuing up for dole handouts at the employment exchange as in the past. Those traumatic memories had never quite left him.

A rather worrying message then arrived from Brenda to the effect that Norman was suffering from heart trouble and had to take life a little easier. Fortunately, he was nearing retirement age so would shortly be released of work responsibilities. Brenda and he were very much looking forward to that event.

A technical development of interest to Matt and his colleagues was the success of the US Pioneer 10 spacecraft in being the first man-made device to escape from the solar system. Also exciting was Matt's acquisition of his first desk calculator, a fairly bulky item but it could add, subtract, multiply and divide with lightning speed, and was a great advance from using slide rules and log tables. (Later on, of course, these new devices could calculate square roots, powers and all sorts of sophisticated transactions.) The immediate effect on Matt was that he could now calculate the combustion temperature of a fuel-oxidant mixture in a matter of twenty minutes instead of the hours that it used to take him laboriously by hand.

This new technology created a great deal of animated discussion in the Engineering Department of the university and, at one of their routine meetings, the members were introduced to Dr Andrew Kendall, an energetic and rather sensitive young man from the Arts Department, who had heard that some lectures on humanities were sought for inclusion in the engineering courses in order to broaden the background knowledge. The meeting then set up a working committee, comprising Andrew and Matt, to investigate the idea and report back. Andrew proved to be pleasantly co-operative, and a humanities course was devised comprising lectures on the history of art, literature and music.

"Now what," queried Matt, "can we offer your students in exchange? After all, it'll be fine for our engineers to be conversant with, say, *King Lear*, but not if your students don't know the difference between Entropy and a bull's foot." Matt made the remark lightly, which masked the strength of his feelings. Andrew smiled, as he was slightly taken aback, not quite anticipating the need for two-way traffic. The thought of 'civilising' these engineering characters, professionals though they may be, seemed eminently worthwhile, but the need for some scientific basis for his own students had not really crystallised in his mind. The two men then repaired to the Senior Common Room to discuss the matter over coffee.

The eventual conclusion was that the humanities course would contain a brief series of lectures on the history of science and engineering, which would also be presented to the arts students. They were both confident that colleagues from the Science Department would be prepared to co-operate, and Matt was certain that he could find lecturers for the history of items like materials manufacture, road building and energy supply. He himself would cover the history of fuel prospecting, winning, refining and use, and also a simplified explanation of engine operation. They therefore agreed to prepare a report for presentation at the next meeting. By this time, the two men had created something of a bond of understanding between them, and Andrew then felt it timely to raise a related query.

"Look Matt, in our department we make a practice of mounting a debate for our students – about once a month – putting some motion or the other in front of them for discussion, and for sorting out the pros and cons. It's a very useful exercise, but I sometimes feel that it's – er, well, a little incestuous since we're all Arts people together. After what we've just talked about, I would like to see some input from the scientific world, and I wonder if you would be prepared to help. I ask because I know you've had a good deal of experience in applied science, and management in the Forces and overseas. What do you think?"

"Sounds interesting, but what would you expect from me?" queried Matt guardedly. "I remember as a student we debated the rather saucy motion, 'That the female configuration is unsuited to trousers.' Can't recall for the life of me whether it was lost or carried, but we young blades had a lot of fun talking about it."

"No, not quite that," smiled Andrew. "What I had in mind was that you might be prepared to write a short piece giving your views on some chosen subject, and then invite debate, possibly suggesting relevant questions that warranted answers. It would be nice if you could be present to hear the debate, but you don't have to if time can't be spared. You would already have served our purpose in prompting the debate."

"Yes, I like the idea of that since I've some controversial thoughts about one or two things, and it would be good all round if they could be aired and discussed," agreed Matt. "Tell you what I'll do. I'll think up some current issues and sketch out my thoughts on them. If you're happy with them, they can form part of your programme. How's that for a scheme?"

"Splendid, splendid. I'll look forward to hearing from you then, Matt. Cheerio."

When Matt next had a chance to sit down and think through the offer he had made, the first topic that entered his mind was related to the fact that Joyce was so sensitive to inhaled smoke, even at low concentrations, and that this had curtailed their social life quite severely. Although he had not smoked himself (apart from a few drags behind the bike sheds as a scruffy schoolboy – and who hasn't?) Matt was saddened to see that so many people didn't seem to realise the damage they were doing to their sensitive lung cells or, if they did, were unable to do anything about it. Clearly, Joyce's situation had made him more aware of this issue than he otherwise might have been, but he would still have looked on tobacco as a drug, and the tobacco companies as drug pushers in view of all their advertising activities. He wondered whether the company bosses encouraged their loved ones to smoke. Or perhaps they didn't have any! In any case, he realised that this practice gave many people pleasure – in fact, in some sad cases it was the only pleasure they had in life – so he didn't want to pontificate or publish scathing articles about it. Hence, restraining his strong thoughts, he adopted a low key approach to his first subject for the Arts debates, as follows:

<u>Comments on</u>

<u>Tobacco</u>

by Dr Matt Gregg

Over the centuries, mankind in different parts of the world has resorted to sucking smoke into his lungs in order to relieve mental stress. In fact, a publication in 1659 extolled the therapeutic virtues of tobacco to such an extent as to threaten the livelihood of doctors! Until recently, the physical side-effects have been recognised only qualitatively, as in "Don't smoke before attempting championship athletics." Modern medicine now confirms the adverse effects of nicotine on the heart and other bodily organs, and that the practice is addictive. It is now also accepted that exhaled smoke represents a hazard to those nearby.

The tobacco industry undoubtedly has considerable expertise in the growing, processing and marketing of a vegetable product, and it would be helpful if no unemployment arose through any changes in the habits of tobacco users. However, for some time now there have been distinct shortages of food in certain parts of the world. More recently, a fuel crisis indicates the wisdom of using biomatter as a perpetual source of carbohydrates for engine fuels, not only as alternative sources but also because of their carbon neutrality, the CO2 produced on combustion being absorbed by the subsequent crop. Debate is therefore suggested regarding the following questions:

1. Should governments make strenuous efforts to dissuade people from smoking, and seek alternative means of relieving stress and, at the same time, ease the load on the health service?

2. Should a policy be adopted by which tobacco companies are required, preferably by international law, to exchange progressively each year a certain proportion of their crops from tobacco to an optimal mix of vegetation for food and for fuel?

When Matt next saw Andrew in the Senior Common Room, he queried him about the debate on tobacco. "It went very well indeed," replied Andrew. "I'm afraid your first question didn't win full support – some students raised issues like 'personal rights' and 'freedom of choice', but your second question really appealed, particularly in the light of widespread hunger and the current fuel crisis. However, the key point is that your thoughtful presentation and critical questions really got everyone thinking in areas that they might not have been aware of before. Our approach was voted unanimously successful, and we would be most grateful if you could suggest some additional topics. Could you manage this, do you think?"

"I'm very pleased to hear it. Certainly, I'll have a go at something else as soon as I can," replied Matt, feeling quite satisfied.

* * * * *

Once again, technical interests were catalysed when the shuttle vehicle Challenger made the first free flight in space but, once again, misery arose over strike action by the miners – this time against pit closures. Again Matt suffered depression at the tragedy of it all, thinking that it would all have to be settled round a table at some time or other, so why the blazes couldn't it be sooner rather than later, and save so much real heartache and bitterness? If coal is no longer an acceptable fuel, why didn't the authorities make that abundantly clear to the general public as well as the miners themselves. But if coal *could* be burnt economically without pollution, why didn't the miners' trade unions locate this fact and broadcast it widely? To Matt, it seemed so much better to seek the most reliable facts and discuss them across a table, rather than acting like spoilt children and inconveniencing everybody, including themselves. It was quite clear that Matt was not conversant with the attitudes and actions of management and unions – nor did he wish to become involved in a world that he felt was not of his choosing. He and politics would never become bedfellows. The miners' strike ended the following year, but left feelings of great bitterness.

One delightful, if somewhat trivial, episode later helped to raise Matt's spirits, and tended to prove his point. It appeared that a certain sweetmeat comprising marshmallow mounted on a biscuit base and coated with a thin skin of chocolate was favoured by airline passengers but, because the atmospheric pressure inside airliners is set at about three quarters that at sea level, the pressure locked in the sweetmeat caused it to expand and expose an unseemly, naked white waistline. A meeting of the biscuit manufacturers and the airline engineers led to a solution whereby the biscuit was made porous so that the pressures equalised and decorum was preserved. As Matt had always said, "When there's a problem, get everybody concerned round the table and sort the perishing matter out once and for all!"

Matt continued to concentrate on his spontaneous ignition work. The first batch of results had been obtained from his pressurised falling-droplet furnace rig, and the student involved had been successful in gaining his PhD degree. Matt and the student together devised some papers outlining their work and that of earlier experimenters. To much satisfaction, not only were the papers accepted for publication in the scientific journals, but a prize was awarded to them both.

The rig itself took up quite a lot of space in one of the test cells, and Matt decided that his next task was to miniaturise it so that it could be packaged as a moderately-sized instrument for use on a bench top. It was at this point that the phone rang in his office. The manager of the staff office at the university was on the line: "May I remind you, Matt, that you are due to retire at the end of next month?"

"Good grief, am I really?" Incredibly, although he had made financial provision for retirement many months before, he had completely put it out of his mind, due to his fascination with the ignition work. He was frustrated to think that unless he could find someone to take over his work, it would just come to a full stop. What a pity that would be.

The university had been very co-operative in offering a set of lectures to advise staff about the new lifestyle they were likely to lead following retirement, but Matt had not bothered to attend since he felt he didn't need anyone to advise him about how to fill his time. His interests in his family and his research were enough to keep him occupied for all his waking hours. However, he could not help but feel emotional when the actual day arrived for his retirement. He had not realised how affected he would be until he discovered, on arriving at his office that Monday morning, that he was still wearing his weekend clothes. Fortunately, there was just time to nip back home and change into his best suit.

Two of Matt's colleagues were retiring at the same time, and all three gave a short speech at the farewell party laid on by the departmental staff. Matt surmised that many people would be delighted at the prospect of the end of the daily grind, and the start of every day being free to do exactly what they wanted. But not Matt. He would miss the stimulus of his regular meetings with colleagues, students and external contacts in industry and research stations, and he felt quite miserable about it all. Joyce made a point of taking him for a walk in the woods, including a visit to their favourite 'Re-enTree'.

A welcome fillip for Matt was the acquisition of his first desk computer. This had a screen no larger than about nine inches, and was, of course, in black and white only, but it gave Matt a hitherto unknown capability of tackling complex calculations, and also of handling word processing much more effectively than with his electric typewriter, even though it had a daisy wheel and an eraser tape. He had been introduced to computing during his survey work in London, but this entailed learning a computer language, in his case Fortran IV, writing out his programmes laboriously on lined paper, and punching out cards before handing them in to a computer department with its roomful of equipment. Now he was his own boss, with the equipment located conveniently on his desk.

A very interesting invitation then arrived for Matt to attend a kart meeting at Kimbolton in order to help with the checking of the fuels used. These class 1 karts were simply-structured four-wheeled vehicles of about 5ft long and fitted with a 100cc two-stroke engine and direct chain drive. They were driven by youngsters ranging in age from about 10 to 16, and could reach average speeds of 60 mph, and up to 70 mph on the home straight. The regulations stipulated that standard petrol must be used, plus a trace of lubricating oil for the two-stroke, with no additions of any boosting materials like aviation petrol, toluene, methanol, aniline or nitromethane. Matt's duty was to help ensure that no cheating of this kind had taken place.

The procedure followed was for the first two karts to finish, and any others that had shown outstanding performance, to be brought to the trackside test area for examination. The engine would be dismantled to check that its dimensions did not exceed the maxima laid down, and a fuel sample was taken for analysis. This proved of a very elemental nature in view of the limited test apparatus available, but the presence of additives could often be sensed quickly, and the sample then earmarked for more thorough investigation later. One of the tests was to sniff the sample to check that the smell was normal, and later that night Matt was reminded of Mike's experience when he found he was troubled with the vision of flashing coloured lights before he could go to sleep. He therefore recommended that any future tests should involve some type of spectroscopic or colorimetric analyser as soon as the cost of these instruments reduced sufficiently to make them available widely.

The family at *High Firs* was now settling in to a new lifestyle when distressing news arrived from Brenda. Norman had had an attack and died. Matt and Joyce instantly packed a couple of cases and went up to Norwich to see what help they could give. Joyce kept Brenda in close company and supported her in the domestic chores, while Matt went round the bungalow searching for any minor repairs that he could make. It was arranged for Brenda to spend some time with her family, who were sorting out the paperwork and funeral arrangements, and afterwards Matt and Joyce felt free to leave, but with a promise to meet up again in the very near future.

Although retired, Matt wanted to maintain his interest in his work, but only on a low-key basis. The first priority was to have more time to spend with Joyce – not just in the home, but with some holidays and gentle travelling, too. Waiting around in airports did not appeal very much to either of them, but relaxing in a coach seemed much more attractive. Matt also had his eyes on either a caravan or a boat so that they could travel at will and stop whenever they wished. After talking it over, they concluded that river cruising would be even more relaxing than towing a caravan – particularly when you had to reverse the confounded thing – so Matt decided he would go and have a word with neighbour Peter, who was a genial ex-navy man with an interest in small river craft. Peter was delighted to see him because he had recently purchased a small boat for use on the local stretch of the Grand Union Canal, and was seeking someone to help him crew it. Matt accepted at once, and the friendship developed as the two nosed their way tentatively between Bletchley and Leighton Buzzard. Once again Matt found himself operating the locks as the canal followed the contours of the land, but the whole pace of life on the water was so much slower that time taken to operate locks did not promote impatience.

Peter soon decided to go upmarket in his boating activities, and purchased a slightly larger boat moored at Bedford for cruising on the Ouse. This boat also had a single engine that was water-cooled from the river and so drained itself as soon as it was switched off. This meant that they could continue to cruise throughout the winter with no danger of coolant freezing and splitting the engine block. Initially they repeatedly traversed the same stretch of the river in order to gain experience, but there was no lack of interest at any time because no two voyages were identical.

It came as a great boost to Matt's morale when the university contacted him and enquired whether he would be prepared to deliver a concise set of lectures on fuel technology each year. It appeared that no replacement lecturer was to be appointed to cover this particular course, hence his services would be greatly appreciated as the most economical step for the university to take, eliminating the need for accommodation, insurance, etc. Matt was delighted, and so glad that he had continued to scan the technical literature after his retirement, because he could then ensure that he was remaining up to date. Since he would have only about three hours of lectures, instead of the original 20, he had to condense his material extensively, and also redesign his visual aids and his handouts, but this proved a very useful discipline.

By this time, public pressure augmented engineers' own concerns regarding the adverse effects of fuel lead additives, both on the health of the human race and the efficiency of catalytic converters fitted to exhaust systems. Unleaded petrol therefore appeared at British forecourt pumps but, unlike continental practice, this fuel was slightly *more* expensive than the conventional leaded 4-star. Fortunately, the car Matt now had was a *Renault 11* which could handle both fuels without engine adjustment, but lists were prepared of those cars that were restricted to the leaded variety because of potential problems of the valves grinding their way into the cast-iron cylinder head. In the expectation that the Chancellor would create a duty differential in favour of unleaded fuel, some service stations installed a recorded warning on the pump to the effect that the petrol about to be bought was unleaded and might not suit the buyer's engine.

Matt was delighted when Peter decided to go even further up market and purchased a handsome 30 ft cruiser fitted with two diesel engines, cabin heater, fridge and drinks cabinet. The main attraction, as far as Matt was concerned, was the outstanding manoeuvrability. This was proved when the rudder cable of a nearby twin-engined boat snapped, but the boat was brought back to the mooring with no difficulty by judicious use of the two throttles.

Several incidents arose during their shake-down cruising, including the rescue of hapless boaters who had stranded themselves on mudbanks. Putting both diesels into full throttle gave more than enough power to drag the boats into deeper water. Fortunately, the provision of a depth sensor in the keel saved Peter and Matt from such indignities.

"They'll never believe me!"

Once one of their fellow boaters found a keepnet full of eels entwined round his propeller. He reported that the net was saved, and the eels were delicious. But the most alarming episode of all was when two crocodiles floated past (yes, crocodiles in deepest Bedfordshire), fortunately (for the boating fraternity) in seemingly dead mode. It transpired that some exotic-pet enthusiast, tired of tending two hungry reptiles, had disposed of them into the river, where a comfortable environment was encountered at the cooling-water outlet of the local power station. When the powers-that-be decided to shut this station down since it was now in excess of requirements, the life-giving outflow of warm water was ended, with the resultant demise of the two unfortunate beasts.

The need to have a fairly extensive repair on the boat, involving the removal of both propellers, led to a somewhat hilarious episode. After repairs were completed, the boat was returned to the water ready for use. Peter and Matt climbed aboard, started the engines, and put the throttles forward ready to move ahead. To their utter amazement, they hit the boat behind them! It dawned on them both instantly that the two contra-rotating propellers had been fitted to the wrong shafts, but having left the berth they felt they had to proceed and then make some sort of return manoeuvre. Much to Peter's credit, he proceeded out into the river, with the throttle levers in the reverse position, and then moved them differentially in the direction opposite to normal, to effect a tight turn and re-enter the marina. Edging the boat into the mooring under such conditions was tricky, but Peter managed it with the help of Matt ashore pulling on the bow and stern ropes. Having vented their frustrations on the marina owner, our heroes then repaired to the nearby hostelry to seek solace.

But perhaps the greatest hazard met by our intrepid pair was at that moment in early Spring when, as they cruised gently past a riverside park occupied by a lone caravan, the door burst open to disgorge no fewer than five attractive ladies who were not only outstandingly well developed but, to an engineer's approximation, some 93% naked – the remaining 7% comprising a smile, a pair of stockings, and a device believed to be known technically as a suspender belt! Peter's and Matt's eyes met (briefly!), and then they gazed in disbelief at this nudefest, wondering what they had done to deserve it. The surprise factor was absolute, with Peter's jaw plunging into the general vicinity of the scuppers, and Matt's eyebrows disappearing into his woolly hat on a seemingly permanent basis. The explanation was provided when a frenetic photographer emerged from the caravan, and proceeded to record for posterity scenes of the girls playing beachball, the remaining scenery – such as it was – and even our two worthies aboard their boat in the background. It was then that the penny dropped, and Matt was able to observe later that he is probably the only academic ever to have appeared in the picture pages of a girlie magazine!

A further adventure arose when Peter and Matt spotted a small outboard-engined craft come roaring out of a boatyard, and turn into the river so tightly that it capsized, throwing the two occupants into the water. Peter instantly selected maximum throttle, and sped towards the luckless crew. A bystander on the bank, not being aware of the drama, shouted at them to slow down, but Peter pointed at the upturned boat and continued speeding. It transpired that the two idiots had been hitting the sauce to excess in the boatyard clubhouse, although Matt found them both cold sober when he fished them aboard. Having landed the soaked pair, Peter and Matt then retrieved the craft and brought that back ashore also. They felt quite pleased with themselves, having saved at least two lives and one craft, but were somewhat miffed later to find not only a complete lack of thanks from the rescued pair, but also that the boatyard was awarded salvage, even though our heroes had actually delivered the craft to shore. However, they philosophically accepted the score as 'Fickle Fate 1; Peter & Matt Nil'.

By this time both Peter and Matt felt they had graduated sufficiently to begin cruising in style, and that it was now appropriate to invite the ladies aboard, so Joyce and Peter's wife were able to accompany their spouses – and provide copious quantities of tea and cakes aboard. Later on, the company of four explored the various hostelries within striking distance of the river banks. By this time, Matt felt confident enough to be able to handle a twin-engined cruiser on the river, and so he started a tour of the boatyards and was very pleased to locate *Pearl*, a rather elderly twin-diesel cruiser, rather less fancy than Peter's, but oozing character and charm from every porthole.

Although Matt and Peter continued their frequent cruises together, it was now possible for Matt and Joyce to invite any of their family members and friends to join them as they explored the river from Bedford to St Ives. Something about the fresh air and the necessary exercise in scrambling about a boat gives rise to a very healthy appetite and a feeling of well-being. It was a pleasure to invite Brenda for the weekend, and give her a day out on their boat. She enjoyed it immensely, not only for the boating experience, but just to be with people, especially friends of long standing. Her family was being very attentive and helpful, she explained, but there were inevitably long periods on her own, and she admitted to switching on the television, even if she was moving about the house, just to have some background conversation and noise going on. Matt and Joyce found this most poignant, and vowed to do more whenever possible.

One of the aspects of the boating scene that attracted Matt's attention was the variety with which boats were named – clever, amusing and/or outrageous. Matt noticed that when married couples are lucky enough to share a boating interest, their choices sometimes reflected this. Newlyweds, for example, tend towards names like *Just You and Me*, or *2's Co*, whereas older hands might settle for such titles as *Ersandmine*, or the very apt *Me-and-er*. Those who aspire to quirky grandeur choose something like *Queue Eat Who*, whereas the more cheeky adopt *Tender to QE2*. Memories of a Swiss holiday can exceed those of spelling if *Idlevice* is anything to go by, although one honest character opted unashamedly for *Beau Nydle*. Some find it difficult to decide, as in the name *Knot Shore*, whereas others can be misleading, as Matt discovered when admiring the pleasant Welsh-like overtones of *Llamedos*, only to realise when noting the reflection in the water that the name was intended to be read backwards! Matt's eyebrows raised when he encountered a craft named *Passing Wind*, but he concluded that the cleverest name of all was the *Sir Osis of the River*.

* * * * *

The Piper Alpha rig fire in the North Sea causing such a loss of life troubled Matt greatly, but he was particularly gratified to be invited to a directorship of a new company being set up by three colleagues to mount short courses for industry on the topic of transport fuels. These courses were very well received and, largely on the strength of them, an advance request came from a government department for the mounting of a worldwide conference at Bath on aviation fuel specifications, to be managed by Matt. During the months while this request was being finalised, Matt made it his duty to visit Bath and tour the accommodation facilities. He was, not surprisingly, made very welcome at all the hotels he visited, and he gradually absorbed a great deal of the history of that city.

As he collected his notes together, it dawned on him that here was a superb chance for him to try once again to prepare a city history in a readable light-touch style. This gave him considerable satisfaction, particularly when he was able to persuade Steve to visit Bath and provide him with a representative photograph. He was delighted also when Joyce proved fit enough to make a visit, and he was able to show her the main places of interest, including the famous Roman baths, the museum of fashion, and the Sally Lunn baker's shop.

Matt's description of Bath can be found in Appendix 4.

The conference proved very popular, and Matt's Bath leaflet was received with pleasure by all the delegates, particularly those from America.

The downside of all this activity was triggered by the fact that, having expertise in typing and shorthand, Matt drifted naturally into the position of clerk for the small company of colleagues. Matt would have preferred to spend such time on preparing and updating his lectures rather than typing letters and envelopes, which could well have been done (and better!) by a part-time secretary. However, the general consensus of opinion was that the company was not yet sufficiently large to carry any extra staff, and Matt began to feel the strain of coping with a delicate wife as well as all these extraneous tasks, interesting though they were. Hence, when one of his colleagues had to resign through ill health, the resulting decision to close the company came as a distinct relief to Matt, and he concentrated even more on household duties and general caring of Joyce. In fact, although he didn't realise it at the time, he was beginning to suffer from the stress of overwork.

By this time it was expedient to take Joyce about in a wheelchair in order to save her the effort of walking any appreciable distances. Matt undertook this duty with alacrity, although he found it almost impossible to take Joyce to the woods because of the sandy soil. He also found it inconvenient to take her through the supermarkets, as the specially adapted trolleys had not been introduced then, but he persevered and was glad to start to relax when they reached the checkout. Still feeling somewhat flustered, Matt was humbled when Joyce turned to him, her face glowing with pleasure, and said:
"Oh, Matt, I *did* enjoy that!"

This reminded him that Joyce spent so much of her time now at home that an outing of this kind was a treat, even though to him it had been a bit of a nightmare. He resolved that he must take her out more often.

Later, Margaret had a quiet word with Matt to the effect that she had noticed that Joyce was slowing down somewhat, and Avril confirmed this when she next visited them at *High Firs*. "Yes, I know what you mean," replied Matt. "I was getting that impression myself recently, but I assumed it was due to that bout of cold she had. Do you think it might be more than that?"
"Frankly, Dad, yes I do," rejoined Avril. "I think a visit to the doctor would be well worthwhile."
"Fair enough. I'll arrange that for her right away."

The doctor advised a much slower lifestyle for Joyce. "After all," as he explained, "she is running virtually on one lung, and has done very well over these last years, but she should now really take things easily."

Matt took over much more of the housework, and sought advice from Joyce regarding cooking the meals. Margaret and Avril came over as often as they could, and Joyce seemed to respond to the care, shown in her more stable health. One day, she suddenly found that her body was covered in a rash of some kind – possibly due to an imbalance in the several tablets she had to take. The doctor visited and prescribed an ointment that solved the problem, but a few days later it became clear that Joyce was quite seriously ill. Avril came over every night to sleep with her while Matt was relegated to the spare bedroom, but eventually it was necessary to take Joyce into Milton Keynes hospital for more intensive care. Avril went with her, and slept that night in the same side ward as her mother. Matt was awakened by a phone call at about two o'clock in the morning. It was Avril. "Dad. I'm ever so sorry to tell you, but Mum died just a few minutes ago!"

Matt was speechless. The rain was streaming down the window pane, as if in sympathy with his own tears. Joyce. His lovely Joyce, was no longer there. How on earth could he go on? How could he live without the companionship and comfort of the lovely girl who was his wife and lover? He felt as if he had been kicked in the stomach. Matt woke Dorothy, who took him downstairs, made tea for them both, and listened quietly to him as he vented his sorrow in a torrent of tears and words. Steve then joined them, having been to the hospital to collect Avril, and they all tried to bring comfort to each other, but the world was changed irrevocably for all of them, and they knew it.

When it came to the funeral, Matt was adamant. Neither of them had been convinced of the existence of some superior being, nor were they happy with the behaviour of many of the clergy. What was the point, thought Matt, of giving some cleric the outline of Joyce's life, only for him to regurgitate it all to the mourners. They just did not need telling about it – they knew all about it by living it. And any pomp and circumstance involved in pallbearers and a high quality coffin was just a waste of money and effort. Also he was not going to appeal for help from the church when they had both deliberately avoided attending it in the past. Agnostics they may have been, but hypocrites they were not.

So Matt arranged for Joyce's remains to be taken alone with dignity to the crematorium in the simplest way possible. The only outward sign of ceremony was to be the placing of five roses on the coffin – one each from Matt, Steve, Avril, Margaret and Mark. At the moment when he guessed that the cremation was taking place, Matt went for a walk by himself through the woods, finishing up by their favourite Re-enTree that they had enjoyed on so many occasions. He rested his head against the tree and quietly spoke, as if to Joyce herself:

"Goodbye, my dearest one. Thank you for being such a lovely wife to me, and such a wonderful mother to our children. I loved you deeply, and… "

His body racked with sobs as the whole sadness of the event overwhelmed him. A hand fell on his shoulder. "It's all right, Dad. It's all right. Come home with me. Come on. Take my arm." It was Steve. He had seen his father leave the house, and followed because of the obvious signs of distress. Steve led the way back, Matt not knowing where he was or what was going to happen. His immediate reaction was that he didn't want to live any more without his beloved Joyce, but gradually time began to heal, and he realised that there were others who still needed him. His family still sought guidance from his experience, his grandchildren loved their Granddad and would be doubly saddened if they lost him as well, and the university had demonstrated a need for his knowledge and particular skill in transmitting it.

After casting about for some sense of direction, Matt was able to harness his inner strength, and discovered that work was the answer, because he was genuinely interested in what he was studying, and it took his mind off the tragedy that had just occurred.

When the undertaker's account arrived, Matt enquired as to what it would have been if he had used all the usual formalities of a conventional funeral. The difference in costs amounted to several hundred pounds. So, having settled the undertaker's account, Matt then sat down and wrote a cheque for that difference and sent it to Cancer Research. Much better, he thought, for that money to go to saving someone else's life than spent on religious rituals. Joyce, he knew for certain, would have approved. He also decided that the same procedure must happen when it was his turn to leave this world.

This thought was reinforced when Dorothy knocked at his door and asked if she could speak to him. "Certainly, Dorothy, and thank you for giving your support to us all through this time, and through all the years."

"I've been only too happy to do so, and I want to thank you for giving me such a happy home here rather than having to live on my own. But what I wanted to say was that I was very impressed with the quiet way that you handled Joyce's funeral, and I want to ask you if you would do the same for me when my turn comes."
"Dorothy, I give you my word that I will. That's a promise."

Soon after, while Matt was walking through the stationery section of a department store, his eye fell on a bereavement card that bore the words:

> The greater the love
> The greater the loss

This matched exactly his own feelings, and he stepped forward to buy it. But then he stopped. Why buy it? After all, those words were now burnt into his brain for ever. He could leave it to comfort someone else.

With the help of Steven and Avril, Matt arranged for a tree in the local crematorium to be dedicated to Joyce, and a plaque to this effect placed at its base.

Chapter 19 – Gain (1990 – 1997)

The downfall of Mrs Thatcher, and replacement by John Major, at the beginning of the final decade of the 20th Century made little impact on Matt in view of his recent bereavement. He was, however, stung by the revelation that Iraqi troops had invaded Kuwait in order, as they put it, to save their brother Arabs. However, news of the sacking of Kuwaiti hospitals, and mistreatment of the patients, gave that the lie, and Matt was enthusiastic for action to be taken. He therefore supported the Gulf War which led to victory the following year, but he was incensed to hear of the deliberate release of crude oil into the sea with all its damaging consequences, and also the sabotaging of the oil fields leading to pollution from dozens of well fires which had to be extinguished by specialist teams of fire fighters, including those of Red Adair. It seemed a pity to Matt that the advancing Allies had to stop at the Iraqi border, since that loathsome regime could have been eliminated once and for all, but you couldn't invade another's country unless they represented a threat to yourselves.

Matt's interest in world events and politics was reduced further by his preoccupation with his fuel studies to help ease the pain of his loss. He recognised clearly the kindness shown by the family in rallying round him, and realised that he would be completely lost without it. Even so, whereas life had hitherto been so full of colour, contentment and sheer happiness, it was now grey and one-dimensional, seeming to consist of one continuous battle against depression, with no relief of the kind that had for years been available from Joyce's companionship, loving support and the warmth of her soft rounded body in his arms at night.

Thinking it over yet again one day, he concluded that marriage had meant so much to him that he yearned for a second chance, although a flash of guilt crossed his mind. Would this mean disloyalty to Joyce and her memory? On the contrary, he reasoned, it would be proof that Joyce had been so successful in keeping him so happy as his marriage partner. But who on earth would look at him twice at his age? After all, he concluded, he was no oil painting, and didn't have the charisma that Mike had. He supposed people in this situation went to a lonely hearts bureau, but he was sceptical about such agencies. Just because two people seemed to be matched on paper didn't mean that they would be in real life. It must be most embarrassing for a couple to meet up like this, when they're all the time assessing each other's every move to see whether a permanent relationship looks promising. Perhaps it would be better, he mused, if six men and six women were introduced to each other by a bureau host at some sort of party, and then left to get on with conversing in a more relaxed way. After two or three such parties, quite a number may well be able to identify suitable partners. If he were asked what sort of new partner he would look for, he would say someone like... er... BRENDA!

It hit him like a lightning strike.

His mind wandered on... 'But of course. She's very attractive in personality and looks. She is 'free', being a widow. And she's been the best of friends always with Joyce, and latterly with me. I wonder if she would be interested, or whether I could persuade her to consider it? Surely she would be better off under my protection than living on her own with only a TV set for company?' The prospect of this happening, however remote, gave Matt a new incentive in life, and he began to tackle the daily tasks with new vigour and enthusiasm.

Matt knew that Brenda's family were churchgoers, and she herself had been somewhat distressed by the lack of a funeral for Joyce, but he felt it possible for the two of them to agree to differ with mutual respect once she understood his outlook on the matter. He therefore started to write to her, expressing his feelings for her with increasing clarity. One of the thoughts driving him was a rising feeling of panic. He had come to realise how attractive she was, and feared that some other man – perhaps a member of her church's congregation – might court her before she was truly aware of his own feelings for her! At first there was no reply, but eventually a letter arrived just as Matt was leaving home for the university.

Matt was due to sit at the back through one lecture given by a colleague, and then follow with a lecture himself. The letter was burning a hole in his pocket, but he didn't want to open it until he could read it quietly on his own. However, impatience got the better of him, and he opened it just as his colleague started to talk. The first words of the letter shattered him. "Thank you for your kind words, and suggestion that we become closer, but I honestly don't feel that I want to marry again!"

It went on to give a more detailed explanation, but Matt simply stuffed it in his pocket, and tried to counteract the black cloud of depression that seemed to encompass him once again. He did his best to concentrate on the colleague's lecture, and then strode to the front of the lecture room to make his own presentation. This was where professionalism took control. For content and steady reasoned delivery, the lecture was probably one of the best he had ever given, although there was not a trace of the usual humour with which he invariably laced his material, and which had always helped to make it so digestible and appreciated. He then excused himself and left for home, where he tried to occupy his mind as widely as he could to mask his deep disappointment. It was only later that he realised that his colleagues had been watching him closely to ensure that he was mentally fit enough to still handle lectures after so short a time since his loss.

The next few weeks were particularly trying for Matt since he felt that his only lifeline had been snatched away, but he plodded on and wrote to Brenda again to try to strengthen their friendship, and to enquire whether she would be prepared to meet him if he drove up to Norwich, on the understanding that he would stay in a nearby hotel overnight. Eventually, Brenda wrote again, this time to say that he would be very welcome to visit and to stay in her bungalow. Once again, Matt's heart rose, and he arranged to visit as soon as possible.

The journey up the A11 this time found Matt with hopeful anticipation of what this meeting might bring. It would be lovely to be able to see Brenda and talk with her, even if nothing else eventuated. Brenda welcomed him warmly enough, and they enjoyed a quiet meal at home, talking over old times. When Matt retired to the spare bedroom, he tried to settle to sleep but found that the thought of Brenda sleeping in the very next room was tantalising in the extreme. So near, and yet so far. The next day they went for a walk by the Yare riverside, and Matt found great calmness in having a companion about whom he felt so strongly. He also realised just how handsome she was, and wondered why he hadn't noticed it before. But then, of course, he'd always had eyes for Joyce alone.

That evening, the conversation came round to the short-course company work that had left him so stressed. And when he described how he had reassured Joyce that she was as pretty as ever, only a few hours before her death, the sobs returned and he once more veered towards the edge of a breakdown. Brenda then realised how traumatised he was, and what a decent, kind individual he had always been. He was not unattractive, either, and was the sort of man you could respect and – yes – come to love. Almost before she realised what she was saying, she said: "I think we ought to go to bed together, and I'll give you what comfort I can."

Matt was speechless with delight, relief, joy, and every other related noun in the thesaurus. This was something he had started to dream of recently, but never dared to hope that would happen. He wasted no time in getting into bed, and cuddling Brenda just as he had done Joyce. Out of respect for Brenda, as had been the case with Joyce all those years before, this was not the time to go any further. They must wait. Matt was content with the sheer delight of being able to cuddle an attractive woman who was already fixed in his mind as a loved one of deep affection. Matt was very reluctant to leave for home after that weekend, but he knew he must, and he left Brenda with the deepest gratitude and the request to visit her again as soon as possible.

Matt's visits to Norwich became so regular that he felt he was wearing a groove in the A11, but his relationship with Brenda became closer and closer, the two of them finding great comfort in their walks together into the city for shopping, and by the riverside to watch the boating. Dinners at the various restaurants proved especially pleasant as they shared conversation, and were able to dip back into their past lives when they had met so frequently and watched their children growing up. They were both known as Uncle and Auntie to each other's children, and their growing relationship was greeted with pleasure by all members of their families.

Eventually, the time came when Matt took Brenda to a very special restaurant dealing with particularly high-class cuisine and, after an outstanding meal, he took Brenda back to her bungalow and proposed to her. He had forgotten much about the art of wooing because it had been such a long time ago with Joyce, but he soldiered on and did his best, even to the extent of getting down on one knee! Fortunately for him, Brenda accepted, not in pity over his obvious need for feminine companionship, but for the fact that she genuinely looked forward to spending the rest of her life with him. Matt was, after all, the sort of man whose arm she would be delighted to hold, and she would be proud to introduce him to her friends.

Being a churchgoer, of course, Brenda wished to have the ceremony in her local church. Matt couldn't have cared less where the event was to be held – he just wanted to have her as his wife and partner – so he accepted her wishes without question. He did, however, make a point of meeting the vicar beforehand and putting his point of view, but was assured that the church was there for use by the public, and that he had every right to be married in it.

On the day itself, the vicar suggested that Matt and Brenda walk up the aisle together rather than following the conventional meeting at the altar, to which they agreed as suiting their already close relationship. Relatives and friends from both sides were well in attendance, happy to see these two long-term friends finding solace with each other after their losses. Both Steve and Geoffrey kept their camera-operating fingers fully employed, and all the females present looked thoroughly attractive as females always do at weddings.

Brenda had a long-term friend who, unfortunately, was in a hospice suffering from cancer. Being unable to attend the wedding, this lady sent along her husband with strict instructions to note important details such as the colour of the wedding dress, and style of the going-away outfit so that she could visualise the occasion from her hospital bed. Such a poignant act added to the overall emotion of the day, and Matt was amazed that he could ever have found such happiness again after all he had been through.

That night proved to be a pinnacle of joy for Matt. He was now married to someone he held very dear, who had known and loved Joyce as much as he had, and with whom he could now consummate his love.

For the honeymoon, Matt and Brenda chose a week's cruise on the Rhine and Moselle. So it was British Rail by Intercity to Liverpool Street, then by taxi to a Swedish hotel nestling in the shadow of Victoria Station, which in addition to the usual facilities included a switch for heating the floor of the en-suite bathroom! There was in-house entertainment provided by a guitarist and a female singer. The 6 foot width of the bed rather puzzled Matt as to which way to lie on it, and to ponder on the Swedish way of night life! Although the food was excellent, they controlled their appetites just in case the Channel crossing decided to be rough.

The coach arrived at Grosvenor Gardens in good time the next morning, and all baggage was stored aboard with the assurance that they would not see it again until they reached their destination on the Rhine. The drive through south-east London passing near Brockley was a nostalgic experience for Matt, but excitement rose as they traversed the A2, then the M2 and finally reached the massive docks at Dover. The 75 minute crossing in the *Pride of Calais* was smooth but misty, and they found the two continental coaches awaiting them at Calais ready to whisk them off to Andernach, some 300 miles away. The journey was eased considerably by the use of motorways and freeways, and they were intrigued to find that they simply drove through the border posts at Belgium, and then Germany, with no hassle or paperwork. They were assured that the Belgians were proud of the fact that their motorway system is equipped with road-lighting throughout.

It was then that they encountered the great mystery of the journey, for their windscreen suddenly needed its wipers despite the fact that there was no rain! All was revealed when they noted the leading coach belching smoke and fuel spray from its rear end. Frantic hoots and flashing lights served their purpose, and a layby manifested itself with brilliant timing. The fault emerged as no more than a split fuel feed hose, but some passengers became unduly alarmed when one of the drivers approached the scene from behind his lighted cigarette. "He'll blow us all up!" was the troubled concern, at which point Matt was able to reassure them that diesel fuel is just not flammable, so there was no need to panic. (Secretly, however, he admitted that this fuel is ignitable, and that spillage onto the hot exhaust pipe replacement was anything but comfortable.) The coach crews tackled the pipe replacement effectively, and thirty minutes later they were all back on the motorway.

Matt had already made some perfunctory attempt to master a little German, but was put off by the discovery that a 'Berliner', which he'd always assumed to be a resident of Berlin was, in fact, a jam doughnut! Although not unduly sensitive, he felt that our word 'exit' was preferable to 'ausfahrt' at the junction signs on leaving the motorways.

It was late at night when they reached their cruise boat at Andernach where, as promised, they found their baggage waiting outside their cabin door. Having absorbed gratefully hot soup and cheese on toast, they crashed out in their bunks. The next morning gave them an opportunity to explore their cruise vessel (Das Boot). Matt noted that the *M V Virginia* was 222 feet in length and 30 feet wide, with a capacity for 125 passengers. She was powered by two 8-cylinder Deutz 1100 kW diesels, which Matt was intrigued to note were controlled automatically unattended, and cruised at 11 mph. Matt could detect little wash to trouble the banks, probably because the two five-bladed propellers rotated counter clockwise.

The name of the Rhine (Das Rhein) apparently stems from the Celtic word for current, which is a healthy 4 mph, the river rising at St Gotthard in Switzerland and flowing through 820 miles to the Netherlands delta into the North Sea, each 100 km distance from its source, being displayed on boards on the banks with typical Teutonic thoroughness. The banks are protected by boulders and show little evidence of erosion. Matt expected that the rule of the river, giving right-hand port-to-port passing would be observed rigidly, however there were occasions when vessels preferred to pass each other starboard-to-starboard. To do this safely, each vessel had a retractable square signboard on its starboard side which was displayed when necessary, and illuminated by a flashing light. At a particularly narrow river bend, a system of light signals on the bank indicated the presence of vessels approaching the bend from the other side.

The Rhine is a busy river, for commercial as well as leisure purposes, and from their deckchairs, Matt and Brenda saw many tankers and freight vessels each day. They noticed with some surprise that the tugs pushed, rather than pulled, their barges, which were all specially designed to fit into each other, rather like pieces of a jigsaw puzzle. The wheel houses were generally arranged to be retractable, no doubt in order to cope with any extra low bridges, and each vessel seemed to be carrying its skipper's car on deck. They saw one freighter loaded with no fewer than 400 cars, and several barges carrying stone mined from the local hills on their way to the Netherlands as a gift to help handle their flooding problems. They were told that the depth of the river is typically 9 ft, but that this is monitored continuously by a system of 'Pegel' houses to warn of flooding.

The banks revealed many vineyards, even on the steepest slopes, and also numerous castles, often perched high up the hillsides. The purpose of the castles was to serve as toll houses so that the 'owner' of that particular stretch of water could exact dues from all passing commercial traffic. With a fairy-tale environment of this nature, legends invariably abounded. Some involved the odd dragon or two, but most were based on robber barons and archbishops on the one hand, and hapless virgins on the other. Matt surmised that the reputed activities of the former would have led to a marked shortage of the latter!

Matt noted that railways ran on both sides of the river, and that they seemed to be in constant use, with very long freight trains on the eastern side (some of them carrying army tanks!), whereas trains on the western side seemed to be used mainly for passengers. The railways frequently tunnelled through the steeply sloping hillside banks, some of the entrances being decorated in the form of castles. This, they were told, was to fool the Allied aircrew during the war who would not be inclined to attack 'just another castle'. Their fears about the Rhine being polluted and smelly were completely unfounded, since good housekeeping procedures in handling waste and sewage had resulted in the return of the salmon.

The daily pattern of events evolved as cruising in the morning, and sightseeing ashore in the afternoon. After a 27-mile cruise upstream, the boat moored at Boppard, a pleasant resort-like township with quaint architecture and walks along the riverside. Matt and Brenda were pleased to stretch their legs ashore and stroll around the streets and square, window shopping and comparing prices with things back home. They quickly fell into the habit of dividing all marked prices by three to give the approximate equivalent in sterling. The next day saw them passing through the wildest section of the river, and the famous Loreley rock where, according to legend, a blonde songstress sat on the cliff top and distracted sailors to their doom in the rapids.

Their next stop, Rudesheim, proved to be another bustling little town with many shopping streets. Matt and Brenda chose the busy Drosselgrasse (Thrush Street) lined on both sides with inns, wine bars and cafes, each of which had its own little orchestra and postage stamp sized dance floor. They found that when the orchestra in their cafe stopped for liquid refreshment, they could still hear and enjoy the music issuing from the cafe opposite, so it proved a continuous performance.

In the evening, they all piled into the Noddy train, a loco-camouflaged tractor tugging half-a-dozen carriages on tyred wheels, negotiating the cobbled streets with lasting impressions on the backsides! But any discomfort was made well worthwhile with their visit to Siegfried's Mechanical Musical Instrument Museum depicting entertainment in a pre-TV era. Many of the machines were operated by means of a rotating drum fitted with either a myriad of pins to operate reeds, or a perforated paper roll to direct air to organ pipes. Most intriguing was the grand pianola on which none other than Mozart himself had been persuaded to play in order to cut the master paper roll. They therefore heard a piece of music played exactly as Mozart had done, and could even see the piano keys moving – a rather eerie experience described by the guide as their 'house ghost'.

Also of interest was a machine playing a violin by means of a small wheel spinning against the strings, the notes being selected by a series of mechanical fingers, and the wheel moving to the appropriate string as required. They were then taken for a wine tasting, and taught that the proper way to assess a wine is by 'slurping', i.e. taking a mouthful and making a sort of reverse whistle by drawing the breath inwards through pursed lips. The few bottles of wine they ordered from the cellars were found soon after, as promised, awaiting collection outside their cabin.

On the fourth day, they turned and cruised downstream for 40 miles as far as the mouth of the Moselle at Koblenz, the name derived from the Roman word for confluence. On the opposite bank above, towered the massive fortress of Ehrenbreitstein. Once ashore, Matt and Brenda visited the well-known fountain, topped by the statue of a small boy which was designed to soak unwary onlookers by squirting water from his mouth every few minutes. No doubt, thought Matt, Koblenz's answer to Belgium's Manekin Pis.

The following day saw them turning into the Moselle and negotiating three large locks. Matt was intrigued to see the lock gates being controlled remotely to slide below the water. After cruising for 32 miles upstream they moored at Cochem, and this time boarded a rather superior Noddy train for a further wine tasting. One of the interesting snippets of information gained was that the Roman soldiers of the area were required to drink one gallon of wine each day to combat malaria! This was an excuse Matt had never encountered before! He also learnt that the Moselle wines were packed in green bottles, and Rhine wines in brown.

Virginia then returned along the Moselle to rejoin the Rhine at Koblenz, and cruise beyond Andernach to Konigswinter, past the seven volcanic hills of Siebengebirge to which the seven dwarfs brought Snow White to hide her from her wicked stepmother. However, Matt and Brenda opted to join a coach tour of the Ahr valley before rejoining the boat at Konigswinter. Unfortunately, the famous Nurburgring motor-racing circuit was not available for coaches that day, but they did see a monastery at which Konrad Adenauer found refuge disguised as a monk after his dismissal by the Nazis from being mayor of Koln, the ruse holding even though Goering owned a hunting lodge nearby. Also pointed out to them were the two hotels that figured in the 1938 crisis – one of which housed Neville Chamberlain, and the other Adolf Hitler, the night before they met and proposed 'Peace in our Time'!

The cruise continued downstream and passed under the bridge connecting Bonn with Bad Godesberg. It emerged that this bridge was badly needed by both communities, and the cost was to be shared equally between them. However, the latter pleaded poverty, so the former loaned them their share of the money in order to complete the project. Bad Godesberg then reneged on repayment, so the good people of Bonn installed a statue on their side of the bridge comprising a man bending down and exposing his bare bottom to their errant neighbours as a sign of contempt.

The mooring that evening was at the twin-towered cathedral city of Koln (Cologne), and Matt and Brenda joined the coach tour which included the famous perfumery of 4711. Apparently, it all began during the historic times of the French Revolution when Cologne was occupied by the French army. On orders by the commanding General that all buildings in the city be numbered consecutively, a mounted quartermaster galloped from house to house and wrote '4711' over the door of the Mulhen's perfumery shop in Bell Lane. The farewell dinner that night was a splendid candle-lit affair, but people retired early in readiness for their morning departure.

The return coach journey took place smoothly, but the first trauma of the day came with the news at Calais that sailings were delayed due to French strike action. Fortunately this applied only to coaches crossing by sea, so Matt and Brenda were able to joint the *Pride of Kent* as foot passengers. They eventually reached their coaches at Dover, and their destination at Victoria, only a little later than planned.

Brenda and Matt had only just left Liverpool Street when trauma number two kicked in with their locomotive breaking down. A diesel loco eventually arrived and eased them into Maryland station where they disembarked and awaited a replacement train. This proved to be the next one scheduled from London, so already had its complement of passengers, but Matt and Brenda squeezed in somehow with all their luggage, and managed to find seats. The crowding eased at each stop, but they found themselves sitting opposite a young lady who asked them, in very fractured English, if Ipswich was Norwich! From difficult conversation, it appeared that she was Turkish, with little knowledge of English, and on her way to a fruit-picking farm at Tunstead. Unless there was transport awaiting her, how she was going to travel the 15 miles or so from Norwich at that time of night was anyone's guess. Matt and Brenda felt that trauma number three was in the offing. Sure enough, at Norwich, a phone call to the farm brought no reply – the phone lines had apparently been damaged by heavy rain. So there was no alternative but to hire a taxi to take them all to Tunstead, and then Matt and Brenda back to theirs. All in all, a very tiring and long day, but nowhere near enough to damage their overall contentment at being together. Matt concluded that 'love is the icing on the cake of life!'

Matt's first task then was to make arrangements for his removal to Brenda's bungalow in the south western area of Norwich (known as the 'Golden Triangle', no doubt by enterprising estate agents), which he felt he could manage using his own car with no need to hire a van, but there were numerous people who would have to be advised of his new address. All this proved to be such an enjoyable task in Matt's grateful frame of mind. Once again, Ozzy was reinstated above his desk, and Matt felt so overwhelmed with quiet relief and deep happiness that he sat down and dashed off the following:

        Love is for Giving – and Forgiving
        Love is for Bearing – and Forbearing
        Love is for Ever – and Forever, Love

He then pinned this up next to Ozzy so that he would continually be reminded of his good fortune. Our Matt could certainly be considered as a 'happy bunny' at this time, and living proof that the darkest night can, in fact, be followed by a sunlit dawn.

Dorothy in particular was very pleased to see Matt happy again. He had always treated her with kindness and respect, and she felt that he deserved Brenda's love. As for herself, she was beginning to feel so tired as to be unable to cope any more with the routine tasks of helping out in the home, and so she approached Matt to enquire about moving into a residential home for the elderly. After some checking around, Matt and Avril located a pleasant establishment near Newport Pagnell, and so Dorothy moved in there with a few personal items of furniture. She settled in well, as she always did with people, and Matt, Brenda and the family were able to make regular visits.

Meanwhile, Matt began to notice that his sleep pattern was disturbed whenever he took alcohol, so he decided that, rather than reduce his already low consumption of the stuff, he might as well go the whole way and dispense with it altogether, concentrating on fruit juices instead. He was going to miss those nutty dark beers, and the glorious apricot brandies, but they just weren't worth paying for with an uncomfortable tummy and a lack of sleep.

Matt then began to make a systematic job of finding his way around Norwich. He found it best to do this on foot, since it gave him time to absorb the landmarks, and to watch the traffic as it negotiated the various lanes and roundabouts. It didn't take too long to learn how to navigate from their bungalow to particular points A, B, C etc, but it took rather longer to figure out how to drive from, say, A to C directly, without returning to the bungalow first. One thing Matt learnt early on is that Norwich is not as flat as Noel Coward's description of Norfolk might suggest. As it was pointed out to him, most cities originated by the side of a potentially useful river and so must, of necessity, occupy ground higher than river level. In fact, Matt found it to be quite hilly in places, and that a good panoramic view of the city could be obtained from the height of Mousehold Heath by looking outwards from the front of the prison. He learnt later that there is an even better view to be gained from the Mottram memorial, but he was already convinced that he could vouch for the legend that the city had a pub for every day of the year, and a church for every Sunday.

Matt was interested to see the remains of the 20 ft high city walls, built of flint and rubble in the 14th Century, and wondered whether they might be saved from further decay by coating in transparent plastic. Most of the fortified gateways and towers seemed to have disappeared, except for the Black Tower, and the Boom Tower on each side of the river to guard against water-borne attack. He began to realise how violent the past must have been if all these defensive arrangements were necessary – in fact, he was to learn much more of this in later years. Returning to the market place at the foot of the castle mound, Matt was delighted to find a stall that was absolutely stacked out with hand tools, nuts, bolts and other consumables at very keen prices, and he vowed that this would be one of his regular haunts whenever he was in the city.

Brenda and Matt paid a visit to the castle, and followed everyone else in dropping pennies through the grating into the keep many feet below. They also visited the dungeons, experiencing complete blackness when the doors were closed, and then walked round the battlements. They were shown the spot where the 57-year-old Robert Kett had been suspended over the wall in chains, dying slowly through starvation and exposure. Visitors were heard to say, "Thank goodness that sort of thing doesn't happen nowadays," but, with memories of Nazi and Japanese atrocities not so many years earlier, Matt said nothing. Their promised visit to Strangers' Hall was most enjoyable, and they were doubly glad that they had seen it because it was closed soon afterwards. One afternoon they also drove the mile or two out to Wymondham to see Kett's Oak. It is heavily supported and looking its age, but it's still there after nearly 500 years since Kett used it as a rallying point for his followers.

Matt's interest in the history of the city also made him aware of the following facts:

* Luke Hansard from Norwich became a master printer in London, and is remembered by the name given to the House of Commons reports;

* Richard Hearne of Mr Pastry fame was born in Norwich;

* The famous Swedish singer, Jenny Lind, gave the proceeds of her concerts in Norwich to establish a children's hospital there;

* Archbishop Matt Parker, a favourite of Henry VIII, was so inquisitive that he became known as 'Nosey Parker'.

When Matt and Brenda visited Brickhill the first time to coincide with Matt's lecture to the university, they found that *High Firs* was once again cat-worthy. A slim black golden-eyed kitten had started life in a Bedfordshire village, but her elderly owner had been taken ill, so the Cats' Protection League had stepped in and re-housed the kitten with Steve and family. Since she was black, and considered invaluable, she was named Penny, short for Penny Black. She was found to be alert, agile and quite fluent in feline language. If only one could understand the lingo, a quite meaningful conversation would have followed. She enjoyed her food, but controlled her intake strictly herself, and so she looked set to retain her agility throughout her life.

An important part of Matt's surveying of Norwich was to locate a boatyard that was reasonably near and had a vacancy. Eventually he found one, and made arrangements to have *Pearl* transported by road from the Ouse to her new mooring on the Yare at Brundall. Normally this would have meant taking the boat through the busy city streets but, by great good fortune, the new A47 bypass was completed and opened on the day before the boat movement was due to take place. Matt soon had her made river-worthy, and so the family was all ready to explore the waterways of the southern region of the Norfolk Broads.

Before they could start this, however, the news came through that Dorothy was becoming very frail. Matt travelled down straight away, and it was clear, even to his non-medical eyes, that Dorothy was fading steadily. She had turned a bilious yellow, and although she could still understand what was said to her, when Matt went to kiss her goodbye on leaving, they both knew that this really was the end for her. A few days later, she passed away, and Matt kept his promise by arranging the simplest possible funeral with no rituals. He also sent a sizeable cheque to Cancer Research in her memory. She had been such a quiet unobtrusive source of support that she was remembered with deep affection by the whole family. Another reason why Matt would never forget her was that her death coincided with his birthday. They arranged for a bush to be planted in her name at the local crematorium.

Matt found that one of the advantages of moving from the Ouse to the Yare was the freedom from negotiating locks. He was also amused to note that the boat in the neighbouring berth was named *Sectimaro* and was told by the owners that this was a portmanteau version of 'Second Time Around', since they had both been married before as, of course, had Matt and Brenda. On the downside, the pump-out loo was no longer permitted, so Matt and his crew had to resume the old routine of 'bucket and chuck it' when they arrived at boatyards with the appropriate facilities.

Cruising upstream from the boatyard led *Pearl* from the Yare to the Wensum, and then past the rail station and Carrow Road football ground to the Riverside Yacht Station where one can moor up to explore the city on foot. As a resident, of course, Matt had no need to do this, but he did take *Pearl* to the end of navigation at Bishop Bridge, and then turn by Pull's ferry to retrace his route. Cruising downstream was rather more attractive in that it gave access to the wider river, banked by copious trees and reeds, with numerous water birds to watch, like the grebe, the coot, and the long-legged heron (known as the harnser in Norfolk). Every now and again one came across a broad, resulting from early extraction of peat, in which it is possible to cruise through slowly and/or moor up. These broads are very shallow, so it is essential to follow the routes marked out with long marker poles and buoys.

Matt and Brenda frequently passed through Reedham, with its chain ferry and swing rail bridge. You have to time your ferry passing carefully in order to avoid snagging your propellers in the chains. They would then follow the river left up towards Great Yarmouth. By turning right at Burgh Castle, it was possible to make a circular tour via St Olaves and the New Cut back to Reedham, and home. On another occasion, they turned off south into the Chet River before Reedham to reach Loddon, where they saw a large tree stump being sculpted into a wooden statue by two artists. The return cruise was somewhat hazardous, because one engine decided to fail. Progress was still reasonable using the other engine and the rudder, but speed was reduced. The problem was subsequently traced to filter blockage caused by water in the fuel and consequent build up of a fungus of micro-organisms, just as Matt had experienced with jet fuels. Matt therefore had both fuel systems cleaned out using a biocide.

Some delightful pubs are to be found at the sides of the river and broads, so it became a routine for the Greggs to invite relations and friends for a day's cruising, incorporating a pub lunch. By this time, Matt felt that his life had reached a peak of happiness again following the trough of despair with the loss of Joyce. It was rather similar to the situation when marrying Joyce following the loss of Mike, but this time there was no world war causing so much death and sadness. Matt's heart was so full that, during one of his walks while Brenda was cooking a meal, he felt the need to try and put his thoughts into rhyme. He recognised that he was no poet, but the urge was there to make some attempt now that he had become more experienced in the use of words. So, over the next few weeks, he turned this idea over in his mind, and eventually came up with the following:

I THINK OF HER

I think of her and all she represents.
Her presence makes the world a lovelier place,
So feminine is she in every sense;
Her dress, her walk, her gestures and her grace.

She shares my moods of gravity and fun,
And lights the spark of yearning and desire.
She slips her hand in mine, and I've begun
My homage as her champion and squire.

The joy of living fades when we're apart
And I'm denied the beauty I adore.
I think of her because she holds my heart.
She is my Brenda. I need say no more.

Matthew Gregg

Well, Poet Laureate he was not, but he had done his best to express his feelings, and was only sorry that he could not have done something similar for Joyce, but he just didn't have the vocabulary then. He rather shyly offered his poem to Brenda, who was surprised but deeply touched to receive it. This was proof, if proof were needed, that she had acquired a very loving husband indeed.

Now that he was free of the short-course company, and in such a happily settled frame of mind, Matt decided that he would embark on a comprehensive textbook to incorporate most of the important items of material he had gathered throughout his career. But he was going to tackle this in a relaxed way, and just keep going gently until it was finished, however long it might take. There was no need for any target date, and it must not interfere with his new-found happiness in any way.

The question of holidays then arose again, and they decided to take two. One was to a warmer climate like southern Spain, where Brenda could sunbathe and swim, and since Wendy had been under the weather for a while, they arranged to take her with them. The second holiday would be for Matt and Brenda to take a week in Eastbourne which they both knew and enjoyed.

A taxi took the three of them to Gatwick, and Matt was amused to find that the small screwdriver and six-inch rule that he always carried in his top pocket set off the alarms as he walked through the security check. He had to explain convincingly that these items were part of his permanent kit, and that he found them so useful for tasks like putting on his shoes and stirring his tea! He was persuasive enough to be permitted to take his seat in the Boeing 737, which soon took off and zoomed up into the wide blue yonder.

Matt was reminded of the last time he flew. This was in an Avro Anson during home leave from Germany just after the war. What had impressed Matt then was that, just as he stood up from his seat in order to reach for a handkerchief, the aircraft decided to fall into an air pocket, and Matt finished up banging his head on the ceiling!

Having reached Faro, they collected their hire car, and Matt drove to their holiday hotel in the small coastal town of Tossa de Mar. This proved to be a delightful spot, and so photogenic with its hilltop castle ruins that Matt spent much of his time exploring the place with his camera while Brenda and Wendy soaked up the sun on the beach.

On their return, Matt got stuck in to some DIY projects starting with some shelving, and then helping a builder friend to install a new fireplace. In the evenings, just for a change, he worked for a while on his new fuel textbook. Brenda was delighted with the home improvements, and they settled down again to a contented lifestyle.

They treated the Eastbourne holiday as a rest cure rather than an activity period, and spent much time listening to the band, and exploring the shops. One in particular comprised a collection of domestic memorabilia, and it came as something of a mild shock to note that many items with which they had been familiar in their youthful years were now considered as historical artefacts. The realisation of ageing usually takes place gradually, but a sharp reminder arises now and again.

Matt shared the general excitement at the opening of the Channel Tunnel. Although he would invariably choose to travel by sea because of his love of the water, he recognised the convenience of being able to entrain, say, in London, and stay aboard until arrival at Paris or wherever. He held great respect for the ability of the civil engineers to drill the tunnel from each end and to meet so accurately in the middle. There had been several aborted attempts years before, but this time they really succeeded.

The next time Matt and Brenda paid their pre-lecture visit to Brickhill they found that *High Firs* now boasted two cats. Penny still retained her role as chatelaine, but a distraught black-and-white kitten had been found clinging to a tree in the garden. Extensive enquiries locally revealed no cat-less households, so the male intruder was promptly added to the ration strength and, in view of its diminutive dimensions relative to Penny, was dutifully named Farthing. Although handsomely marked with symmetrical patches of pristine white on his black background, this feline member of the family enjoyed a sluggish observation of life unfolding around him, being interested only in food and sleep. Consequently he grew apace and soon outdistanced Penny in terms of size. It seemed a little odd that a Farthing was now so much larger than a Penny, but that's life for you.

Matt's next set of lectures on fuel technology went exceptionally well, and Matt realised that he was steadily improving his presentation by being able to home in on the really important aspects, only commenting on the peripherals in passing. He sat in the Visiting Lecturers' Office for a while to read up on the latest course literature of the university. There was a knock on the door, and Andrew poked his head round and asked, "Have you a moment or two to talk, Matt?"
"Certainly, my dear chap. Do come in."

"First, I wanted to say how deeply sorry I was to hear about Joyce. We haven't met since then, but I heard the sad news from colleagues. I'm so very glad that you are able to find happiness again, and I'd like to wish you well with Brenda. I'm also very pleased that you can carry on with your studies – and to give us the benefit of them, too."

"Thanks a lot, Andrew. Yes, it was terrible to lose Joyce, but Brenda has really saved my sanity – and my life – and it's wonderful to think that both of us knew Joyce and loved her so much. It gives us an even stronger bond, and with our families."

"Since you're back in harness again, so to speak, I was wondering about the possibilities of further discussion projects for our Arts Department, if you can manage it, that is." Matt agreed willingly, and as soon as he could find the time, he prepared the following:

<u>Comments on</u>

<u>Hunting</u>

by Dr Matt Gregg

Throughout the centuries, there appear to have been four reasons why mankind has hunted wild animals to death:

* For food. I have never visited an abattoir, nor have any wish to do so, but I eat meat; not just because I enjoy it, but because the anthropologists tell us that our teeth are designed for that purpose.

* For protection. If attacked by a wild animal capable of killing, or severely maiming people nearby or myself, or if it wantonly killed farmed animals not for food purposes, or if it carried a dangerous disease, I would be prepared to kill it if there was no way in which it could be restrained.

* For culling. If it could be proved beyond reasonable doubt that a wild or managed herd needed to be culled to avoid them all suffering death by starvation, I would reluctantly agree to it,

but would prefer not to be involved. I understand that this happens at a nearby deer sanctuary, but it is done in low key without people dressing up in fancy clothes and classing it as a sport to be enjoyed.

* For blood sport. I was once shown in a market town an iron ring fixed to the ground that used to be used years ago to tie a bear to be baited by dogs. Apparently, some bears had their claws and teeth ripped out so that the dogs had more chance to inflict injury. Recently, I witnessed a man carrying a hare in his arms, and leading a dog. He then released the hare and, seconds later, the dog, which overtook the hare and tore it to pieces. I found this disgusting, even if it is called sport. Later I was shown a cage full of birds and told that these were 'for the guns'. There was no way that they represented a population excess, a hazard to other species, or even a source of food.

Debate is therefore suggested regarding the following questions:

1. Is there medical evidence to show that meat is an essential component for a healthy diet?

2. Should we be prepared to kill animals only for food, defence or the proven need to cull?

3. Now that much of the world seems to have moved on from bear baiting and, hopefully, cock fighting, should we aim to move still further by recognising blood sports as cruel, and discouraging them accordingly?

4. Should we encourage those who enjoy blood sports to be honest and admit it, so that they can be granted more respect by so doing?

Matt thought that that ought to put the cat amongst the pigeons! He expected there might be some strong opinions expressed in the debate, but that was exactly what he was trying to achieve, in the hope of leading to better understanding all round. It also helped him to get his own ideas sorted.

In the municipal world, the problem of disposing of garbage persisted, and interest was growing in the use of landfill gas for local power stations. The Brogborough plant near Brickhill was a typical example. But what caught Matt's eye was a reference in an Australian Institution journal to the effect that, in the city of Whittlesea, seventeen plastic pipe wells had been drilled 7 metres into an old landfill site, and connected to an automated plant room where the gases were compressed for use in a barbecue located in a suburban park. What with that and the ice-vending machines, picnickers out there were really well catered for.

A great source of contentment for Matt was the continuation of the university to invite him to present his lecture course in fuel technology, so he and Brenda timed their visits to Brickhill to suit. The fact that he was interested enough to monitor the literature on the subject meant that he kept up to date, and so could not only present the latest in this particular discipline, but could handle the students' questions with greater confidence and authority. These invitations now spread from the degree courses to the short courses that the university mounted for members of industry, so Matt was able to rub shoulders with practising specialist engineers as well as upcoming students.

Each short course concluded with a formal dinner, and both Matt and Brenda were invited to attend. It was traditional for a humorous speech to be given on these occasions, usually by Professor Taylor himself, who was no mean stand-up comedian, so the courses ended on a particularly pleasant note. However, other speakers were also invited from time to time, and Matt was delighted when he was given the chance to make an after-dinner speech. He was, of course, well versed in addressing an audience, and in lacing his address with humour wherever appropriate, but he did not consider himself to be a comedian, or even a humorist despite his attempts to write humour rather than speak it. However, he did his best, as shown below:

Professor Taylor, Ladies and Gentlemen,

First I wish to thank you, Professor, for inviting us to this event, and for giving me the opportunity to speak tonight. As we all know, Professor Taylor is someone who is never ruffled by surprises or crises, whatever they may be, and can always handle people with tact and courtesy. He would have made a splendid contribution to the diplomatic corps but, fortunately for us, he opted for the engineering profession instead. If I can remind us of the definition of a Diplomat, he is someone who, when he says Yes, he means Perhaps; when he says Perhaps, he means No; and when he says No, he is no diplomat!

[A ripple of laughter went round the room.]

I believe there is a codicil to that regarding the definition of a Lady as someone who, when she says No, she means Perhaps; when she says Perhaps, she means Yes; and when she says Yes, she is no lady!

I don't quite understand this, (said Matt in mock innocence), but perhaps someone could explain it to me later in the bar.

Next I would like to thank Freda, the Professor's secretary, for arranging this occasion, and also for looking after every one of us so thoroughly during our working lives here, particularly for shielding us from the demands of our students. In fact, we could say that Freda has been a mother to us all.

… Incidentally, Freda asked me not to say that, so I won't.

You know, giving an after-dinner speech is not the easiest of tasks to perform. Sir Winston Churchill used to say that it is like trying to climb a wall that's leaning towards you, or kiss a girl who's leaning away from you. I'm doing my best, but I'm not very good at climbing walls!

When I graduated from university all those years ago, there was none of the unemployment problems that we have today. In fact, I was given a nice blue suit to wear, and instructed to invade the continent of Europe. This was all part of World War II – you must have heard about it – it was in all the papers. Apparently, the fact of my enlistment was somehow relayed by spies to the German High Command, and no less a person than Hitler himself was heard to say, "Pilot Officer Matthew Who?" You can't have a greater accolade than that! I can't say that my contribution to Hitler's downfall was outstanding in any way, but I did manage to survive the war years without ever catching my spurs in my chin strap, so that's something to be quietly proud of.

As you know, I have spent some thirty years in this education business, man and boy, and back then we didn't have the electronic office equipment that we enjoy today. In fact, the only software we had was a female secretary, and I'm not convinced (glancing at Freda), that we have really made much of an advancement since then."

[Simulated frown from Brenda. Giggle from Freda.]

We also had a chance to move to Australia, a land of much sunshine, sand and, surprisingly, snow. I recall on arriving there saying to myself, 'If I make a success of things here, I'll stay permanently.' Well, folks, it's good to be back! But I did learn a few things during our stay. First, I was taught how to throw a boomerang – and to duck when it returned, to save having my teeth knocked out. Also, I was shown how to cook a galah when out in the bush. Now a galah is a highly coloured sort of cockatoo – quite attractive in appearance but, unfortunately, approaching plague populations. First, you build a fire and then put some water in your billy can. Incidentally, a billy can must never be washed from the day it is made, no matter what it may have contained since. You hang the billy can on the fire, and put in the galah, together with the steel head of your axe. Galahs are very tough birds indeed, and you only know they are cooked when the head of your axe goes soft!

We now live in Norwich, which has been described in the literature, rightly, as 'a fine city'. I was somewhat surprised to note that, despite all the transport and movement that takes place in this country, local dialects and accents still endure. For instance, here in Bedfordshire people still say, 'Wait while Tuesday', rather than, 'Wait until Tuesday', whereas in Norwich you will be asked, 'Are you rate?' rather than 'Are you right?' You'll also hear words like 'bootiful', 'compooter' and 'fooneral'.

There is a large square castle that sits dramatically on top of a man-made mound, with an ancient market place located at its foot. Because any public insurrections generally started in the market, it was located in its present position so that the authorities could keep an eye on things. There's an interesting pub there that is unique in that it has two names. Originally it was known as the Gardener's Arms, but it appears that one dark night the owner did away with his wife in the cellar, so the name was duly changed to the Murderer's Arms. Eventually, the city fathers decided that this name did not quite suit the image of a respectable city so they ordered a return to the original name. This incensed the clientele, and all the drinkers rose as one, in defiance of this ruling. This tense situation was resolved in the good old British way of compromise by making the sign bear one name on one side, and the other on the reverse. The drinkers then retired contentedly to their tankards.

Mind you, Norwich caters for a wide variety of tastes, and nearby can be found an establishment geared specifically to tired businessmen. Apparently the main feature there is lap dancing – whatever that is – probably something like musical chairs.

Unfortunately, the hostesses are so poorly paid that they cannot afford tops to their dresses. Perhaps they should seek assistance from the IMF, or the Lottery. Incidentally, the only reason I know all this is that I've seen it on Anglia Television – well, that's my story, anyway.

I recommend you visit Norwich, where you'll hear all sorts of stories, mostly connected with the agricultural scene of Norfolk. Typical of these is the one I was told where a farmer made his routine monthly visit to the livestock market in Norwich, and bought himself a cockerel. On his way back to the station, he passed a cinema that was showing a film he dearly wanted to see. Since he didn't want to wait for another month, he slipped the cockerel down his trousers, and paid to go in. All went well for a while until it became clear that the cockerel was stifling in the cramped space, so the farmer could do no more than open his fly and let the cockerel's head poke out to breathe. Again, all went well until an elderly lady sitting nearby noticed something unusual in the dim light, and whispered to her sister next to her, "Hey, Doris, this man here is exposing himself in public!" "Lucy, my Dear," replied Doris calmly, "Don't be alarmed. When you've seen one, you've seen them all." "Well, this one's different," replied Lucy, "it's just eaten one of my crisps!"

Now this, ladies and gentlemen, is the sort of robust humour that you'll encounter in that part of the world. But if you do visit, and happen to meet me in the lap dancing establishment, please be kind enough not to recognise me – this will save a lot of questions in the house!

Well, ladies and gentlemen, the years go by, and there is no way we can stop them. I am reminded of that American commentator who said that there are three ages of man:

1. He believes in Father Christmas.
2. He doesn't believe in Father Christmas
3. He is Father Christmas.

In my case, I find I have had to give up skateboarding, which is a pity. But I can assure all you younger people that medical advances today are such as to offer the elderly much greater comfort and activity in their lifestyles. Mind you, there can be some side-effects that must be watched. For instance, would you credit it, but that wretched Viagra stuff is giving me dandruff? I mean, moi?" and he stroked his bald head.

Talking of age, I am indebted to a Professor in Brighton for this classic example of graffito. Apparently an advertisement in Sussex promotes cross-Channel travel by proclaiming 'Newhaven for the Continent'. Underneath has now been added the words, 'Eastbourne for the incontinent'!

In conclusion, may I let you into a secret that even my colleagues have never known after all these years, and that is that this moustache of mine that you see here – is a fake!

…Yes, a fake (repeated Matt). I keep my real moustache in the top pocket of this jacket.

Ladies and Gentlemen. Brenda and I thank you sincerely for the fabulous food, wonderful wine and, above all, this charming company that we have enjoyed tonight. Thank you so much.

Matt sat down to widespread applause, and a proud kiss from Brenda.

* * * * *

Of particular interest to Matt were the developments in the field of measuring anti-knock quality of motor fuels. Matt knew only too well of the constraints involved in octane rating with the splendid, but expensive and sensitive, variable-compression engine, each test taking an hour or two of work. Now, however, with the use of techniques such as infrared spectroscopy, a fuel sample could be analysed within seconds, and automatically compared with the data from several hundred samples previously tested and locked in the instrument's memory. Matt began to think that this technique might eventually lead to the measurement of many other properties of fuels, and that several current standard test methods may steadily become redundant.

As regards testing for diesel fuels, a bench test apparatus was now marketed that involved spraying a fuel sample into a heated chamber under pressure in order to determine the ignition delay, and hence the cetane number, very much on the lines that Matt had envisaged with his own extensive rig just at the time that he had to retire. He was delighted to see that such a rig had arrived, but only sorry that he had not been able to take an active part in its development.

Meanwhile, back in Norwich, things were moving apace. A disastrous fire had occurred in the library, apparently due to an electrical fault, and many valuable items were destroyed although a team of experts was able to refurbish many other volumes. This gave an opportunity for a new library incorporated in a stunning new building, to be known as the Forum, designed with a complete glass frontage and containing restaurants, a museum and eventually a BBC studio. An imaginative design of shopping mall was constructed within the mound of the castle, incorporating cafes and a multi-screen cinema. The old Norfolk and Norwich Hospital was decommissioned by stages and relocated in a new building complex at nearby Colney (pronounced 'Coney').

Holiday time came round again, and Brenda recommended an 8-day coach tour of the Isle of Wight, where, she said, she had never felt more relaxed and care-free. Matt, who had not visited there before, agreed eagerly. So once more they found themselves aboard a feeder coach in Norwich which took them to the interchange at South Mimms, and then via the holiday coach to cross the Solent from Portsmouth to Fishbourne, finishing up at Sandown.

Apparently, the Isle of Wight is smaller than Greater London, and was occupied on several occasions by the Romans, who knew it as Vectis. Its relatively recent discovery stems from the 19th Century when it was recognised by the Victorians, who developed it as a holiday retreat. Queen Victoria herself occupied Osborne House, and Brenda was particularly intrigued to find that a wardrobe door in the Queen's bedroom led directly into the bathroom!

Overall, the attractions lie in the mild climate, the sweeping bays of Brightstone, Freshwater, Sandown and Alum, the wild headlands of Culver Cliff and the Needles, together with the lush chines at Shanklin, Luccombe and Chilton. Also the 12 different shades of coloured sand at Alum Bay which were formed 50 million years ago under the sea, have given rise to a multitude of souvenirs. Sandown itself lies at the heart of Sandown Bay, occupying six miles of sloping beach backed by low red cliffs, and the island enjoys one of the best sunshine records in the UK. Carisbrooke Castle proved intriguing, particularly because of the fact that Charles I was imprisoned there as an honoured guest, although he did try to escape, just before being taken to London for execution.

Two items were of special interest to Matt. The first was the fact that the island experiences a double high tide. This is because the tail end of the rising tide flowing up the Channel turns round the eastern end of the island and boosts the tide starting to ebb in Southampton Water. The second item of interest was the leafy Shanklin Chine with its remaining 65-yard length of pipe-line that gravity fed Bambi, the Sandown-Shanklin pumping station of PLUTO (Pipe-Line Under The Ocean). This ingenious project for a direct supply of fuel to the Allied invasion forces in Europe arose from the question, "Could you run a pipe-line under the Channel to supply oil when we invade?" posed by Lord Louis Mountbatten, as Head of Combined Operations, to Geoffrey Lloyd, Britain's Secretary for Petroleum.

The first lines, described as HAIS (Hartley-Anglo Iranian-Siemens) were based on submarine telegraphy cable technology, but the later versions, known as HAMEL (Hammick-Ellis) were of simpler flexible steel pipe design with a mere 6-week life expectancy. Eventually, four 70-mile lines were laid from Shanklin pier to Cherbourg, and seventeen 29-mile lines from Dungeness to Boulogne. Before a depth-charge test could be conducted to check the vulnerability of the line, a German bomb dropped nearby, effectively demonstrated resistance to damage. To preserve secrecy, many precautions were taken, including covering the Bambi pumping stations with a false roof loaded with either rubble from bombed buildings or grassed earth, sweeping away the tyre tracks after each vehicle movement, and regular aerial photographs by the RAF to check the camouflage. On reaching Europe, the pipe-line system was extended across the continent as the Allies advanced into Germany, and a total of 350 million gallons were transported at pressures up to 440 lb/sq. in. After the war, the pipe-lines proved a nuisance to shipping and the GPO's cross-Channel cables, so were salvaged, some of which were said to have been made into boiler tubes for railway engines exported to Argentina.

Matt, who had of course seen the receiving end of the pipe-line project in Europe during the invasion, took great delight in recounting these facts to Brenda, and was so enthusiastic about it that she found it all of interest, despite a limited concern with matters technological.

The balance of exceptionally attractive scenery, and this fairly recent technological history, served them both with a thoroughly enjoyable holiday break, but this pleasure was shattered on the journey home by the horrific news of Princess Diana's death in a car accident in Paris. The wave of grief that swept the country was outstanding, and Dulcie insisted on following hundreds of others by standing on one of the bridges over the M1, and throwing a bunch of flowers onto the hearse as it drove up to Althorp.

Chapter 20 – Viewpoints (1998 – 2000)

During one of his visits to the university, Matt was once more approached by Andrew who asked if he would consider yet another discussion subject. He explained that these discussions proved extremely useful, but they only had a limited time for them, and sometimes they were a little slow to get going. This was where Matt's notes came in so handy because they gave some mature background and then offered some suggestions to kick-start the proceedings.

"Well," replied Matt, "I do have some fairly strong feelings on some of the issues of the day, but I'm not sure whether they'd suit your purpose. For instance, animal rights come to mind. I have a strong affection for animals – as you know, I'm a lifelong member of Redwings Animal Sanctuary in Norfolk for example – and I do feel we have a responsibility to care for all animals, farmed and wild, since we are the ones gifted with the brains to do so. And I'm not happy about laboratory animals in medical research, although I do understand the need to test new drugs somehow before they're released for human use. The sooner the researchers find some alternative way of testing, the better. And certainly, animals should never be used to test cosmetics, in my view.

"It's really disappointing to find such an important issue treated so clumsily. After all, there must be thousands of people out there who feel as I do, and we ought to be encouraged to band together in a responsible way instead of being outraged by the sort of terrorist tactics that we're all so sick of. This is a classical case of the pen being mightier than the sword. Wouldn't it be better to come up with a series of punchy articles outlining alternative means of checking drugs to combat diseases? With thousands of us in support, this would create a research pressure that could not be ignored by government and big business. After all, the problem is not going to be solved by people rushing about with placards. It's the medics who will eventually solve it with some blob of biomatter or clever electronic device, so we should encourage and fund them rather than fight them. Once it's solved for good, we'll all feel much more comfortable.

"Still, I don't know whether there'd be enough difference in opinions amongst your students to make for a good discussion. But perhaps I could give it some more thought and come back to you later. Would you like me to?"

"Yes, please do that. Now what about the delicate subject of homosexuality, Matt? How do you approach this with your scientific background?"

"Well, I think that also lends itself to a modicum of common sense. The key factor is that men and women are designed and programmed to fall in love with each other, and to create children for the next generation. As you know, I feel strongly that this should happen in wedlock rather than before, but I'm well aware that this makes me a dinosaur in this promiscuous age when everything, from coffee to sex, has to be instant. But sometimes Nature gets her wires crossed – perhaps 'knickers in a twist' might be more appropriate. Forgive me, Andrew, I'm not trying to be crude. In these cases, males attract males, and females attract females. I look at this as an accident of Nature – I do *not* agree that this is just another way of living, and feel that if something can be done by hormonal rebalancing or something, then it should be done.

"I well recall a boy at our school telling us that he had six toes on one foot! Of course, we told him to stop talking rubbish, but he then whipped off his shoe and sock and, sure enough, he had a sixth toe growing out of one of his ordinary ones. This was clearly a mistake on the part of Nature, and I believe it was corrected later by surgery, so if homosexuality can be corrected medically, then I think it should.

"I do support the idea that homosexuals should be accepted by society without the criminal overtones in cases like Oscar Wilde and Alan Turing. I mean, just think of the benefits that people like them and Noel Coward have brought us. But I think the idea of homosexuals 'marrying' is utterly unacceptable. They should come up with some other term, as they have done in prostituting the word 'gay'. Marriage is involved with the potential of parenthood, and even if two homosexuals bring up a child, as well as the case of single parents, I think it is a pity that the child does not have the benefit of guidance from both a man and a woman, since this would help in handling friendships and relationships later in life. One thing that did disgust me was when one homosexual dared to suggest that his way of life was actually cleaner and more worthy than a conventional heterosexual one – that is arrant rubbish. Incidentally, I would not readily seek advice from a homosexual clergyman – if he can't sort his own emotions out, how on Earth could he help me? But, once again, I don't quite see this as a suitable discussion topic, do you?"

"No, perhaps not, Matt. Any thoughts on royalty and republicanism?"
"Yes indeed. We've had a pretty turbulent history in this country regarding plots and strife to gain the throne, but I believe it has all enabled us to settle down now and achieve an optimal balance between Crown and Parliament. You know my critical views of politics and politicians, and I shudder to think what would happen if we lost the overall steadying influence of the Crown. Like it or not, prime ministers would edge their way towards dictatorship – we've already seen examples of that sort of arrogance recently. It doesn't matter too much, in my view, if the reigning monarch or family do behave in some questionable manner – although our present Queen is an example of duty and decorum. What matters is the continuing presence of the Crown, which the politicians must recognise and defer to under some predetermined arrangement. I strongly advocate that those who prattle on about republicanism bear this in mind when they urge us to delete the royal family. Anyway, that's another of those political issues which is not quite what I thought you had in mind, Andrew. So what topics can we bandy about now? There must be something. Let's see."

"Well, there was a problem recently about evicting travellers from a field near the university. Any thoughts on that, Matt?"
"Certainly, I think the solution lies in their own hands. I have every sympathy for people wanting to up sticks and live a roaming life – I sometimes get attracted to it myself. But if these people occupy an area uninvited, and then leave behind a mass of old tyres, burnt out cars and general rubbish, they can't expect anything but resentment from those of us who live settled lives. If only they had the sense to adopt a policy of leaving a site even cleaner than when they arrived, that problem would be resolved at a stroke. The only other thing, of course, is that everyone, including them,

using the road system must pay the appropriate taxes. If they did all that, they would find themselves much more acceptable to society. So it's up to them."

"What about immigration, Matt? That's also featured in politics, of course, but aren't most things?"
"Yes, but that also lends itself to a straightforward common sense approach. The first point is that the native British population is steadily rising because of much reduced infant mortality rates, and the extension of life for the elderly. But Britain is a small island, unlike America or Australia, and we already have evidence of the loss of countryside, even greenbelts, and pressure being put on supplies of water etc. We've already seen, for example, that Newport Pagnell is now virtually contiguous with Milton Keynes, and there's talk of siting hundreds of new houses in this area. So we *must* keep an accurate measure of how many immigrants are arriving. And since we have a scheme of immigrant admission, anyone abusing this should be sent back immediately.

"Secondly, there are far too many people in the world subjected to repression for reasons of race, politics, religion or whatever, and these should be our first priority from the viewpoint of sheer compassion. After that, of course, we should welcome skilled and professional people, but I have some worries on this score. Certainly these people would be assets to this country, but what about their own countries? Wouldn't they be needed even more back there? Are we being selfish in creaming off the better educated from, say, third world countries? Rather than keep pumping funds into these countries, wouldn't it be better to train up the most likely candidates and send them back to strengthen their own cultures and so enable them to become independent of our handouts? Are they going to be dependent on the so-called rich countries for ever, or are they going to accept our help to become truly independent?

"One very useful contribution we could make to this end would be to organise an instruction course outlining the various systems of government and civil organisations, including our own, and then encouraging them to adopt their favoured system in their own country so that they can develop their economy and stand fully on their own feet. Until that time, I think we should insist that any financial aid given should be fully audited so that there is no possibility of improper use – as has certainly happened in the past. So controlling immigration is a benign combination of common sense and compassion, and it disgusts me to hear politicians criticising this as racist. It's nothing of the kind. But then, I say again, you know what I think about politics, Andrew. Anyway, what's the point of holding a census now and again if we can't keep track of the number of immigrants?"

"Yes, I do. Well, what about astrology?"
"Yes, that would be a good topic to explore, but I think a research approach might be better than a discussion. For instance, the students could construct a table listing all the zodiac periods down the left-hand column, and along the top write in as many sources of horoscopes from magazines etc that they can find. Then, every week or whatever, they could complete the table, although they would probably have to learn to summarise because of shortage of space. If they then studied their summaries, they could see if there was any uniformity in the predictions and, if so, they could enter these in the right-hand column. Mind you, to do the job properly from Go to Whoa, they could construct a similar table with zodiac periods on the left, but along the top put names of people who are prepared to volunteer, reporting any happenings of note during the week. Once again, any uniformity in the summaries could be listed on the right. The crunch would come when they compared the right-hand columns from the two tables. If they matched up, this would be a terrific shot in the arm for the astrologists, but if they didn't, it would be clear that the whole thing is just a tongue-in-cheek pastime – hopefully harmless, although I believe that some historical celebrities have relied on such predictions before deciding to invade or whatever.

"Of course, if astrologists really believe what they tell us, they would welcome a study like this. So if they resent it, then we can be pretty sure that it's all hokum. I would love to see this study done, but don't suppose it ever will be. Not long ago, some students – particularly in the politics departments – were very active in revolting against the establishment. What they lacked in experience they made up for in passion. But somehow I can't see even these revolting students," Matt smiled at the pun, "tackling this project. Pity, really, because it would teach them how to record information in an impartial, unbiased way, and how to summarise honestly without changing the thrust of the argument, just as we do in scientific research."

"Matt, I think you've got it. What about suggesting something on the Scientific Method?" Matt replied to the effect that this is mainly a matter of applied honesty and common sense in a search for the most probable truth, however unpalatable this may appear to be. "But, to be fair," he added, "it's not reserved entirely to science. Where it does not apply is to politics and religion."

"Oh, do you really think so?" rejoined Andrew. Although not strongly political, he was a committed Christian, and Matt's remark rather stung him.

"Indeed I do," replied Matt. "There are only two things that I think we can take for granted. One is death, which comes to all of us, commoner or king, and must do so otherwise we would take up space and resources for the following generations. The other is that no one, now or in the past, has ever really known whether or not there is some supreme being. If people wish to think otherwise, then I'm happy for them and am glad it gives them comfort, but I say that they have no right to indoctrinate young minds into accepting only their own beliefs. This is just brainwashing, and has been practised for centuries. No wonder religions have endured for so long. I know for a fact that I was constrained to one religion only, and frowned upon when I asked questions about it, whereas in every other subject at school we were expected to question, and were castigated if we didn't. How about you, Andrew? Were you ever given any instruction on any other religion than that of your parents?"

"Well no, not a lot, but we were assured that this was the most benign and acceptable of all of them."
"Of course you were. This is the way it works."
"Yes, but the clergy feel they are doing the right thing for their flock. In any case, don't you feel that there must be some supreme being, whether you call it God or whatever?"
"I honestly don't know, Andrew, but I see no indisputable evidence. Wouldn't the world have gone on just as it does without? Someone once described religion as 'institutional wishful thinking', and I believe he was

right. Show me some real evidence, and I'll start believing. This is the main difference between the clergy and people like me. They say they know for sure that there is a God, whereas we say we don't know for sure that there isn't, but remain with an open mind to be convinced. It doesn't help to find that so many hundreds of different faiths and beliefs exist, with such conflict between them. You know, in science, the natural laws that are discovered are generally accepted by everyone, irrespective of race or gender."

Andrew interrupted him at this point. "Look, Matt, you've clearly got some strong feelings about all this, and I want to hear them, so why don't you let me take you to lunch? After all, you've served our department well in the past and we've never really compensated you for it. So what about it? We can carry on our discussion there. My car's just outside."
"You're on," replied Matt. "I accept with pleasure, as the saying is. But I'll just ring Brenda first and tell her I'll be home late."

A few minutes later Matt was seating himself in Andrew's car. "I see you have a Japanese car, Andrew. Are you happy with it?"
"Yes, I am. I've tried other makes, and some are very comfortable, but I'm told that the reliability of these cars is outstanding."
"You're quite right," replied Matt. "I have one myself and it gives next to no trouble. I remember after the war I decided not to have any more contact with the Japs over their treatment of POWs, but you can't carry a hate for ever, and I'm glad that we're now able to put all this behind us and make a new start. Then, when I tried to buy a new English car, I couldn't because of a strike. I was so angry about it, that I went straight out and bought a Japanese car."

They drew up at the pub, and Matt recognised it as the same hostelry where Mike and he had spent their last convivial evening together. Although still in a village environment, known as Middleton (i.e. Middletown), it was now surrounded by the massive development of the new town of Milton Keynes. It gave Matt something of a nostalgic feeling as he settled with his orange juice at the same table he had occupied all those years ago, and waited for the scampi lunch.

"Coming back to our discussion, Matt, are there any aspects of the church that you do agree with?"

"Certainly there are. For instance, I go for honesty and kindness when dealing with others, and I believe, as the church does, in reserving sex until marriage. I also like the idea of people being able to have access to a kindly, sympathetic ear when they are in any trouble."

"What do you think we should do if there is a God?"

"Pray to him, of course, but the way people pray really horrifies me."

"In what way, Matt?"

"Well, it seems to me that people who pray are always asking for something – like 'give us', 'forgive us', 'lead us', 'deliver us', or whatever, but on no occasion do they ever add the word 'Please'. This seems to me so bad mannered and selfish. Haven't they already been given so much by their Almighty, like life itself, and health and all the pleasures of living?"

"Mmm. An interesting point, Matt, but I do see that you're very sincere about it. What else don't you like about religion?"

"First, the fact that there are hundreds of different religions around the globe that all started at different times, and some were imposed on people through wars or aggressive evangelism. If there really is a god, then surely we would all be influenced by him, and there would be only one religion? That I would respect, and no doubt follow. If anyone was to criticise the present system by suggesting that each religion must be man-made, and all seem to be fighting for supremacy rather than co-operation, how on Earth can I oppose this view?

"I am no student of history, but we've all spent some time learning about our turbulent past, and it's noticeable that religious differences figure time and again. The gunpowder plot, for one thing, was an attempt to replace a Protestant monarch with a Catholic one, wasn't it, so it doesn't impress me when leading clerics suggest that religion is not a primary cause of conflict. Of course it is. Religion should stand for truth. We do things so differently in science. It's good to have people come up with fresh ideas, even if they sound barmy, but then colleagues don't hound them, or excommunicate them, or put bombs in their cars. What we do is publish the ideas so that they can be considered critically by our peers and then discussed in a mature, professional way."

"Well, all this is pretty strong meat, Matt. What other objections do you have?"

"I'm unhappy with many clerics' self-importance. I realise that, years ago, the clergy were virtually the only learned members of society – that's why our academic dress today is modelled on their robes – so they exercised control over many aspects of life, secular as well as religious. I notice, for instance, in all the villages around here that the largest house was reserved for the vicar – and his servants – whereas the poor old doctor had to make do with a converted living room. Nowadays, of course, all these mansions are sold off for offices and things, but it just shows how much clout the clergy had in those days. Anyway, the world has moved on since then, and there are plenty of people around today who are far better educated and experienced, so I think that the clergy should recognise this, and take a lower profile. I find it very irritating to hear them pontificating on such things as dress, diet, and even circumcision – surely these are the provinces of clothes designers, dieticians and doctors.

"The churches have also got a pretty grisly past about which they ought to be honest enough to admit and seek forgiveness. This wretched business of sacrifices, for instance, whether of animals or humans. It just doesn't even begin to make sense, and looks more like organised mindless cruelty. Then there's poor old Galileo. He had the brains to figure out the relative movement of the Sun and the Earth, and also the guts to proclaim it. But what happened to him? Excommunication, wasn't it – and it took centuries for the churches to admit their error. That wasn't very confidence-giving, was it, by people who profess to be our leaders in thought? I'm also unimpressed by people like monks and nuns who closet themselves away from the real world and spend their time praying, and sometimes even punishing themselves. Somehow this seems so self-indulgent. Surely it would be better for them to do something useful in the outside world, like nursing, or helping out with chores, or even cleaning the drains! This is why I respect the Salvation Army so highly. I don't agree with their religious thoughts, of course, but I do admire the way they care for down-and-outs, drunks and druggies, and are prepared to enter all sorts of dodgy dives to do their good work. That is practical Christianity, and I slip coins into their collecting boxes whenever I get the chance.

"Then there's the question of Creationism. By all means let its followers put their point of view to us, and give us all the evidence they have collected. This is the stuff of learning, after all. But when they insist that only their philosophy should be presented at schools, and the logical science of archaeology be banned, then I think they shoot themselves in the foot, and show how ridiculous they are. Knowledge should be disseminated, not suppressed."

"I find this very enlightening, Matt, although I can't necessarily go along with you all the way. But, before you continue, how about a dessert? I see treacle tart is on the menu. Would that suit?"
"Champion," replied Matt who was thoroughly enjoying himself by being able to make his points of view clear to an intelligent, even though unconvinced, listener.

"If I might continue on that score, Andrew, there's the matter of the Turin Shroud. I was told that this was reputed to be an image of Christ's face on the cloth wrapped round him, and was printed by some flash of atomic energy or something. My instant reaction at that time was that it could not be so. This came to me, not by some divine inspiration but from an understanding of simple geometry. If you look straight at my face now, Andrew – I hope it doesn't put you off your dessert – you will see that that it is about 9 inches high and 5 inches wide, that is a height-width ratio of 1:8. But if a cloth was wrapped round my head from ear to ear and scalp to chin, the ratio would be more like 1:1. In other words, any imprint of my features on the cloth would be distorted, whereas the portrait of Christ was not. But when scientists wanted to check a sample of cloth for its age, the believers declined – almost as if they were afraid of the result. If I were a believer, I would be so confident and keen to have evidence of its authenticity that I would insist on such a test. Anyway, the test went ahead eventually, as you know, and it was proven to be much, much younger than 2,000 years, but I'm sure that some believers will not accept this."

"Once again, I'm impressed with what you say, Matt, but I still remain a Christian for all that. Now what about some coffee? You really deserve it after all you've talked about with me."

"Yes please, and make it a large one if you will. It's been thirsty work, but so enjoyable."

"Right, Matt, coffee is on its way. Have you any more beefs about religion that you'd like to air? I must say that although I don't accept all you say, and I certainly have my own views that I'd like to discuss with you another time – after all, you're not the only one with opinions."

"OK Andrew, I take your point, but I'll add just one more. I've already complained about the arrogance of people believing that their religion is the only true one, and all others are faulty. I get so angry when people come knocking on our door expecting us to ignore everything we've learnt over the years, even though we may be much older, and better educated, than they are, and then try to sweet-talk us into their own beliefs. This is not just bad manners – it's ignorance and arrogance combined, and I really think we should devise a law against it. By all means, let these people put some literature in our letter box so that we can study at our leisure, and put it in the kitchen boiler if we feel like it. I hate being accosted, especially on my own doorstep. And this is where I worry about teaching. I feel that in science we do our best to open young minds, whereas the churches seem to confine them."

"I think," said Andrew, "that it's probably a matter of need. You have your own logic of science and self discipline that give you your strength and philosophy of life, but many people don't and so need a pattern of thought to cling on to – something like a bright light shining through a fog."

"I agree with that, especially with the word 'discipline', but what a pity that there are so many lights, and that we are forced, not guided, to follow only one. You know, nobody says to a child, 'You've got to be a physicist whether you like it or not, so don't have any truck with those chemists, and ignore everything the engineers and medics say, even fight them when you have a chance.' There was no coercion for me to study engineering – I chose to myself. And it's usual for members of the scientific community to pool their knowledge, with mutual respect, for the common good. When a cleric tells me that the churches behave the same way, I'll listen to him. But I'm still searching for that single, universal light in the fog, and have the germ of an idea at the back of my mind. Just give me some time and I'll try and sort it.

"Incidentally, I would have hoped that a marriage of two people of different faiths would have been seen as a heaven-sent opportunity to link religions closer, but I've often found this to cause irreparable strife and even break-up within families. I just don't understand that.

"Now, I think the time has come, Andrew, to thank you sincerely – not only for this enjoyable lunch, but for listening to me so patiently while I've been sounding off. I'm sure that this is the sort of thing we all ought to be able to do without getting upset with each other. You know, I can't talk like this with Brenda because she is so entrenched in her church. Not that this creates any problems between us since we've both got enough sense to respect each others' feelings, and just not raise the subject. I really am lucky to have found her after what's happened in the past, and if I were a praying man, she would figure strongly in my prayers."

"Have you every prayed, Matt?"
"I wouldn't call it praying, but I sometimes feel so deeply grateful for the good things in my life that I talk out loud when I'm walking in the woods. I know that if I did pray, the words, 'Thank you' would figure repeatedly in what I said."
 "Would you be prepared to write down a sort of pseudo-prayer and show it to me, so that I can understand better what you have in mind?"
"Yes, I will do that, and for the reason you say."
"I do see how very sincere you are in your views, Matt. I'm sure you would agree that it would be too fundamental to suit one of our discussion groups, but don't you think you ought to write your thoughts down and offer them for comment somewhere?"
"Perhaps, but I don't want to start firing off criticisms to newspapers and things," replied Matt. "I'm happy to leave everyone to his or her beliefs, but I am tempted to tackle this idea of mine about bringing religions closer together, and then try to find a wider audience. In other words, I want to

make a positive contribution, rather than a negative one. Anyway, thanks again, Andrew."

Having polished off their coffees, they started back to the university.

"I wonder," asked Andrew, "if you could answer a couple of questions that have been bugging me for some time."
"Go ahead. I'll do the best I can."
"Well, first, what is the difference between engineering and technology?"

"The way I look at it," replied Matt after some thought, "is that engineering is the bigger picture, and technology is one of the specialisations lying within it. It's rather like one country lying within a larger atlas. For instance, my topic is that of fuel technology, which is all part of the overall energy engineering including things like wind energy, solar energy, and the geothermal energy of the natural hot-water springs. But, you know, I think the best definition I've seen so far is that from the *Engineers Australia Journal* that goes like this: The function of science is to know; of technology, to know how to; and of engineering to do.

"That reminds me. You know how keen I am on teaching, and how I think it requires a certain talent to do it properly? Well, I've always been stung by Bernard Shaw's trite comment that: Those who can, do, and those who can't, teach. I'd like to modify that last line to: 'and those who can't teach make a hell of a mess of it.' But I'm sorry, Andrew. You had a second question for me. What was it?"

"That's right, Matt. Can you tell me what exactly *is* entropy? I've heard it mentioned several times."
"Entropy? Well it's an abstract quantity, and a bit difficult to explain, but briefly it represents a measure of disorder. Mathematically, it's easy. If we have a chunk of hot material at, say, temperature T(hot), located in a cold environment at T(cold), then we would expect the transfer of a quantity of heat, say Q, from the hot chunk to the cold environment. The loss of entropy of the chunk is given simply by Q/T(hot), and the gain in entropy of the environment by Q/T(cold). So, in this case:

$$\text{Overall change in entropy} = \text{entropy gain} - \text{entropy loss} = Q/T(\text{cold}) - Q/T(\text{hot})$$

and this, of course, is positive because T(hot) is greater than T(cold). In fact, the entropy of the universe is rising all the time. As far as it is of practical use, well, it gives us an idea of how much energy from the hot material is available to be converted into work because, unfortunately, not all heat can be converted to work, although all work can be converted to heat, as with friction for example. You see, work is ordered energy transfer but heat is not, and as Nature prefers disorder to order, the entropy rises. There's a lot more to it than that, of course, but perhaps you get the broad idea."

"Just about, but I'd have to really study it hard to get a proper handle on it."

"You're right, Andrew. Most of us have to live with it for a while to get the full strength of the argument."

"Thanks. I don't know about you, Matt, but I often think that this term 'professional' is being degraded by being over used."

"I certainly agree. To me, a profession involves extensive periods of study and experience, all under the mantles of integrity and responsibility. It angers me to hear talk of 'professional criminal', for instance. Crime is not a profession, for God's sake! They should say something like 'an habitual criminal'. Also the 'oldest profession in the world', should be described as a trade, not a profession. I'm in no sense criticising prostitutes by this, you understand. In fact, they serve the community better than many people realise, in my view, because some men are so highly sexed that they can't cope with only a normal relationship based on love. Without the safety valve of paid sex, there would be innumerable cases of rape around the world, and no woman would be safe. That said, I feel sorry for these ladies who can't find a more rewarding lifestyle, and even sorrier when they have to rely on someone else to organise their activities. Life never ceases to be interesting, does it?"

Having reached the university, they shook hands in a classic example of two responsible minds agreeing to differ, but with mutual respect.

Possibly as an offshoot of these contributions to the Arts course, the students became aware of Matt's willingness to make thought-provoking contributions to their university lives. Consequently, it was no great surprise while Matt was reading an article in a technical journal, that a student knocked on and poked his head round the office door with the following request.

"Morning, Doc. Mind if I have a word?"

"Have several, my friend. What can I do for you?"

"Well, you see, I'm an Australian student, and I've been hijacked to run the student magazine here. But it's bloody difficult to get any copy out of these Poms. Now back home we have a scheme where the *Engineers Australia Journal* asks certain senior engineers a range of questions to gain a snapshot of their thoughts on life."

"Yes, I know. I'm a member of that Institution, and so read the *Journal*."

"Oh, of course. Well, would you mind answering a few questions on the same basis? I have a list of them here if you'd care to see them."

"Certainly. Leave it with me and I'll return it to you as soon as I can."

"Many thanks, Doc. Much obliged."

As soon as he could, Matt settled down and came up with the following:

<u>Snapshot</u>

Q: What do you think is the most important discovery of mankind?
A: Anaesthetics.

Q: What was one of the most exciting projects you've worked on?
A: Operation Overlord during World War II, but I always enjoy passing on knowledge gained to upcoming students.

Q: What made you choose Engineering as a career?
A: Because I was initially fascinated with aircraft, and then the fuel that powers them.

Q: What was one of the best professional decisions you've made?
A: Working overseas to gain a broader view of life… In Australia!

Q: What was one of the worst professional decisions you've made?
A: Not contacting oriental companies to offer an advisory service in English style and spelling in their advertisements. I could have made strong contacts with them.

Q: What has been one of your biggest challenges?
A: Heading up a mechanical engineering department in an Australian university, and also being initially responsible there for the civil and electrical departments as well.

Q: What was one of your best experiences in life?
A: Being happily involved in marriage and family.

Q: What was one of your worst experiences in life?
A: Losing my first wife through tuberculosis.

Q: What makes you laugh?
A: Good British comedy like The Two Ronnies, Les Dawson, Eric & Ernie, Tommy Cooper, Peter Sellers and Peter Ustinov, and also the old Ealing comedy films, and the 'end of the pier' shows at Cromer.

Q: What annoys you?
A: Thick-headed arrogance, brainwashing, torture, crime, political cant, racism, cruelty to animals, and lack of recognition of women's abilities.

Q: What inspires you?
A: Good leadership, honesty, decency, self discipline, gallantry, wisdom and old-fashioned common sense.

Q: What cheers you up when you are down?
A: The affection that engulfs me from my family, my interest in my work, and the respect shown me by my colleagues.

Q: What is one of your strengths?
A: Determination.

Q: What is one of your weaknesses?
A: Insufficient imagination to understand certain abstruse concepts.

Q: What do you do to keep fit?
A: Walk rather than ride whenever I can, watch the intake and quality of my food, refrain from smoking always, keep off alcohol most of the time, and never, never touch drugs.

Q: What do you do to relax?
A: Read widely, enjoy the tranquillity of travelling on water, wander idly in the woodlands,and dabble in trying to write interesting prose.

Q: What is the first thing you do when you get home from work?
A: Renew my contacts with my wife and family, and our two cats.

Q: What do you do on Sunday mornings?
A: Lie in a little, then go for a walk in the woods with my wife when she is free of the cooking.

Q: What are your favourite animals?
A: Donkeys, but closely followed by penguins and meerkats, and, of course, our two moggies at home.

* * * * *

It was at this time that Matt and Brenda were closely involved with Phyllis, Brenda's cousin, who lived in one of the remoter parts of Norfolk. As an elderly lady, she had developed breast cancer, and needed to undergo a course of radiotherapy in Norwich. In view of her age, the obvious solution was for her to stay at the Greggs rather than be transported backwards and forwards from home to hospital so regularly. Phyllis was a compact little person with a very direct manner of speech, so there was no chance of misunderstanding what she had in mind. For all that, she was kindly and humorous, so Matt and Brenda shouldered that responsibility without question, and were as relieved as she was when she was found to be cured.

In the meantime, excitement was steadily growing over the imminent arrival of the twenty-first Century. Many scare stories abounded about computerised equipment failing at the onset of year 2000, even to the extent of aircraft being put at risk in the air but, throughout the usual festivities on New Year's Eve, there was little report of anything dire happening.

The family had managed to gather and welcome the new millennium with a party. After it was over, Matt and Brenda sat for a while on the settee, and reminisced over their meeting, their marriage, and the events of the past few years. "Well, my love, we seem to have reached a milestone in history as we sit here and wait while the new century rolls over on us."
"Do you think there will be all those computer problems that they try to frighten us with?" asked Brenda.
"I honestly don't know, love. I'm not a computer expert myself, but I gather from Steve and Geoffrey that there should be nothing to worry about."
"Well, let's hope they're right."

Inevitably the conversation turned to their families. "I think we've a right to be proud of all our children, and grandchildren," opined Matt, "and how lucky we've all been not having had any accidents at birth, and only minor ones during their lives. I'm so pleased that the youngest ones have turned out as well as they have done. It worried me with this modern idea of poor discipline, so the youngsters are not checked when they start going the wrong way, and at times they've behaved like little stinkers. It's only natural that kids will experiment with their behaviour, but in our day we had the parents and teachers to define the boundaries for us, and to make their presence felt if we ignored them. I remember Dad telling us that he had to give both Mike and me a good hiding when we were being more than usually obnoxious. Not that we remembered that episode at all – it didn't give us any lasting complexes whatever – but it enabled us both to grow up into responsible citizens, and to love Dad into the bargain. Do you recall any discipline dished out by your Dad, Brenda?"

"Oh, yes. He once slapped me on the backside when I'd been playing up instead of going to sleep, but fortunately I'd had the foresight to slip a magazine into my pyjamas beforehand so it didn't affect me greatly. What a little rotter I must have been. Also, like you, I adored my Dad. Yes. I felt that Mark was much too lenient with Janet but, of course, she's part of your family, not mine. In any case, you can't interfere, and she's now turned out to be a very caring person because, I think, of all the pleasant people around her."

"You know, it's not only in child rearing where people automatically take you to mean the other extreme if you try to offer advice. It's like this," Matt sketched on a paper napkin this simple diagram:

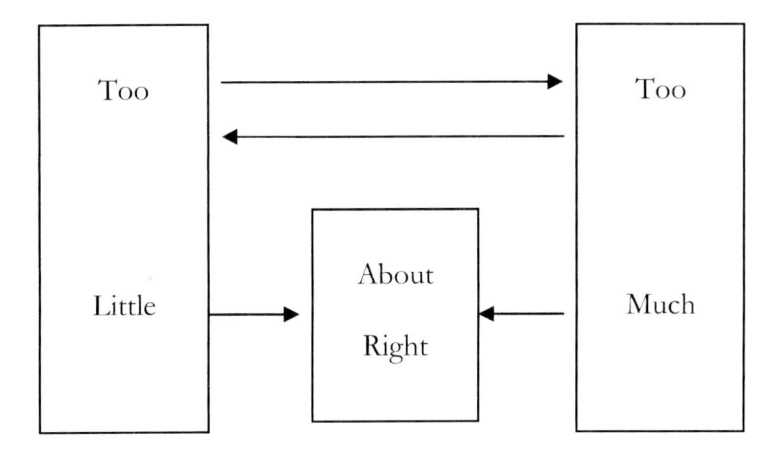

```
┌─────────────┐                          ┌─────────────┐
│             │  ──────────────────▶    │             │
│    Too      │  ◀──────────────────    │    Too      │
│             │         ┌─────────┐     │             │
│             │         │  About  │     │             │
│   Little    │  ─────▶ │         │ ◀── │   Much      │
│             │         │  Right  │     │             │
│             │         └─────────┘     │             │
└─────────────┘                          └─────────────┘
```

"We're not trying to suggest that people swing from one extreme to the other – only from one extreme to a reasonable compromise, but it's darned difficult to get through to some people. When we suggest, for instance, that there's too little discipline, the chairborne do-gooders will instantly accuse us of advocating the other extreme with draconian repression and child abuse when all we're trying to suggest is the halfway mark of a balanced approach.

"Do you remember when we first saw this after the war, when the permissive age dawned, and we knew then that problems would arise? And so they did. There were nothing like the levels of vandalism and crime then that we have today. And the 'no smacking' brigade just cannot and will not see the connection. No, we don't want cruelty to children any more than they do, but we do think that it's kind – and necessary at times – to provide some guidance and show that we mean business… by means of a slap if needs be!"

"I certainly do. Norman and I treated our two the same way, and we always thought that the results showed we were right. And you're right, it's far less safe for a woman to go out alone these days – especially when it's dark."

"Ah yes. I remember when I was a lad visiting Gran and Granddad in Battersea, and a neighbour telling me that in her younger years it was unsafe to venture round Clapham Junction because of the footpads, and we both felt comfortable and relieved that it had now become safe. Well, of course, we're right back to square one now with muggers about ready to rob you to feed their drug addictions. Not exactly progress, is it?"

"True, but I suppose we have to try and keep it in perspective. Fortunately none of us has suffered in this way – so far, anyway – so we must be grateful."

"Indeed. By the way, did you notice with Dulcie that she learnt more quickly at school than did Geoffrey? We found that both Val and Avril were quicker in the early stages. Someone said that this was due to an inbuilt urgency in girls to mature and have a family. I don't know whether that's true, but I do believe it's best for girls and boys to be taught separately. Although I've been involved more with older students, I think that the separated youngsters concentrate more on their studies than on each other. They can learn to mix better outside the classroom. I also think that our system of smaller schools years ago was better since the headmaster could get to know almost all of us personally – and keep us in order better – whereas a huge comprehensive may be more financially efficient on paper, but it's too impersonal to oversee and control behaviour. Still, I suppose I'd be classed as a dinosaur now. I know this, that I would love to be invited to give a talk on fuels and engines and things at a school, but only if I was assured that the class would behave properly, and that I could sling them out if they didn't. I would not tolerate bad behaviour at all. End of sermon," added Matt, smiling.

"I've never taught, apart from giving a bit of instruction in hospital, but I do see the point in what you say. Anyway, it's not our problem now, is it? We are the 'old folk' now, and I don't suppose anyone is interested in what we say about all this."

"I'm afraid you're right, love. Anyway, I'm so grateful for meeting up with you as I did. We've known each other for a long time now, haven't we? I always looked on you as a sort of sister during our first marriages, and it seems to me so right and proper that we should come together eventually. I can't thank you enough for agreeing to marry me."

"Well, you had to work hard to persuade me," said Brenda, smiling, "but I'm so glad now that you did. You're a good man, Matt, and I've come to love you deeply. I never thought you'd have such affection for me."
"Thank you for that, love. And for you."

They didn't say much after that. There was no need because the two minds had fused together, and there really wasn't much more to be said.

Dulcie cleared up the remains of the New Year's Eve party, and while Brenda prepared for bed, Matt experienced one of those rare moments when emotion transcends logic, and dashed off the following thoughts that had exercised his mind for some time, stimulated by Andrew's request:

Matt's 'Prayer'

> Thank you for my life and my health.
>
> Thank you for my family who created and nurtured me.
>
> Thank you for my wonderful marriages to Joyce and to Brenda.
>
> Thank you for all the love I have been able to give, and to receive.
>
> Thank you for our fine children, for their health and for the progress they are making in the world.
>
> Thank you for the friendships that have brought so much enjoyment to my life.
>
> Thank you for my interests and my successes.
>
> Thank you for the strength that has enabled me to handle my failures.
>
> I am deeply grateful for all this, and endeavour to pass my happiness on to others.

Matt then slipped a copy into an envelope addressed to Andrew, and followed Brenda into the bedroom.

Chapter 21 – Unknown (2000 – 04)

Matt studied himself carefully in the mirror as he shaved. This was not narcissism in any shape or form: he would frequently check that age had not made him so ugly that he should no longer inflict himself on a lecture audience. All in all, despite the bald head, the roomy bags under the eyes, and a face with a rather lived-in look, he concluded that, as a strapping young fellow of eighty plus, he could just about get by for a few more years yet. Meanwhile, the twenty-first century continued to take its first tentative breaths of life.

While the western world settled down to change all the dates on its letter headings to 2000, and erase the 199- they'd been using for the last ten years, Matt eventually completed his major text book on Fuel Technology. He had the good fortune to team up with an excellent graphic artist who took Matt's line drawings, and some of Steve's photographs of fields of biocrops, and upgraded them to suit publication. As a bonus, he introduced colour into the illustrations, and even tinted the pages of the different chapters. He also laid out the complete book with its proper pagination, and downloaded the whole thing onto a compact disk. The result from the printers was a massive tome of some 450 pages, and Matt felt particularly satisfied that he had found a way of presenting the bulk of his knowledge in a reasonably compact and readable form. However well it sold, it would serve Matt himself as an extremely convenient source of lecture material, and visual aids.

As the months went by, Matt recognised a bit of a problem. It was true that about half of his book dealt with the principles of chemistry, thermodynamics and combustion, which hardly changed over the years, but the remainder covered all the latest developments that Matt could discover. These, of course, do change continually with time, but Matt could not contemplate rewriting his book every year. Then the solution struck him. Since he was already preparing abstracts of relevant technical articles, as they appeared in the press, in order to update his lectures, he explored the possibility of publishing these as well and, fortunately for him, he found an outlet whereby the abstracts were put on line and made available for fee-paying clients.

* * * * *

The Gregg family shared the shock of the world with the news of the destruction of the twin towers in New York on the infamous '9/11' day with the loss of so many lives. It beggars belief that such hatred exists to deliberately sacrifice innocent lives in this way. This meant that extreme precautions would have to be made in screening airline passengers in future, which would all add to costs and frustration. One's faith in mankind was restored to some extent, at least locally, by the opening of the Friendship Bridge across the Wensum in East Norwich. This pedestrian-cyclist swing bridge was named after one of Norwich's twin towns, the Yugoslav City of Novi Sad.

These dramatic events prompted Matt to tackle Andrew's suggestion of getting his thoughts down in print about his universal approach to the teaching of religion. Although it was way out of his specialist study, he could not help but feel that this might be a useful point to ease the obvious conflict that has existed between religions throughout the ages, even though some clerics argued that the underlying reasons were cultural differences or greed, with religion only secondary. So, having seen Brenda off for her swimming session with Sally, one of her friends in the village, Matt settled himself at his desk once more and proceeded to write. The result was as follows:

A Suggested Universal Route to Religions

by Dr Matt Gregg

Most human activities, once initiated, develop with time as social attitudes change, and this has no doubt always been so with the many religions that have emerged down the ages. More recently, many of the myths and superstitions involved with religions have given way to honest, clear-minded answers to honest, clear-minded questioning, and with them came the birth of Science. Rainbows are but one example: to the primitive mind these were no doubt viewed as signs of the supernatural, but we now know their basis of refraction, and the two angles at which they can be seen. When these rational approaches to the physical world were applied to real life, there emerged the discipline of Applied Science, particularly in the forms of Medicine and Engineering.

Sadly, there has grown some controversy between the apparently blind faith of the religious and the cautious questioning of the scientists, the former believing that they have already discovered Truth, and the latter continually searching for it by challenging their own discoveries. Sadder still is the fact that controversies also appear between individual religions, and even between members of the same religion. To the independent thinker, this is surprising because religious teaching generally exhorts that an approach to the Almighty through prayer can lead to the solution of most problems. The fact that sometimes it does not work is often put down to the fact that the prayer may not have been offered in the proper way. Since the clergy are experts in the art of praying, one could expect that they would be able to obtain heavenly guidance successfully themselves. However, we find deep schisms within certain churches over important issues such as integration with other religions, the acceptance of women clergy, and homosexuality.

There appear to be two separate but parallel routes to help solve this dilemma:

1. The religious could get their act together by explaining clearly why an Almighty who is capable of creating the whole, massive, universe is incapable of convincing everyone on one of its many planets that he (she?) alone is the only such being, and that the religion based entirely on him is absolutely unique. If this explanation is not forthcoming, it would be no fault of

any reasonable, honest and sensible human to wonder whether every one of the hundreds of existing religions is man-made, and that conflict is inevitable between them where their boundaries meet because they are all built on bases that are unable to demonstrate any convincing evidence.

2. The scientists could make greater effort to broadcast the inherent honesty in their methods of deduction by showing how, having assembled a theory devised from a number of facts obtained by unbiased observation and thought, they then attack this theory themselves in order to expose any weaknesses in their argument. Once found, the revised theory is itself attacked, and so on, since the driving passion of the scientist is to find the most probable truth. Anything less than this lays the way open to be selective in considering likely evidence. Some scientists cannot understand why religious people do not follow a similar procedure, since this, surely, would strengthen their beliefs and attract others to follow their lead.

One of the biggest bones of contention, particularly for the scientist who is involved in teaching, is the apparent practice of the churches to indoctrinate their young. In the sciences, subjects are taught on the basis that this is what mankind has learnt so far, but we continue to question and research. In religion, on the other hand, we are often told not to question and, if we do, are accused of heresy. In the West, young people are generally free to choose their career path, marriage partner, area of the world in which to live, and political persuasion, but not their religion. What they need is guidance towards making mature choices. What they don't need is indoctrination or brain washing of any kind, which they will instinctively and increasingly tend to reject.

In the majority of schools, it is noteworthy that pupils attend classes in most subjects all together, irrespective of creed. In some schools, however, the treatment of religious instruction entails the clergy of the different denominations teaching their groups separately in different classrooms. Is this the best way to engender tolerance and co-operation from an early age? Do the clergy feel that it is preferable to ensure that these newcomers are indoctrinated in one, their own religion only, rather than in the fuller brotherhood and sisterhood of the human race? Alternatively, would it not be practicable to devise a syllabus comprising doctrines that are common to all the co-operating religions, leaving the specifics to be covered by the clergy themselves on their sabbath or equivalent days, as shown in the following tables comprising examples of religions that are all fictitious but based on representative existing doctrines:

Table 1 Random Doctrine Arrangement

Fictitious Religion A	Fictitious Religion B	Fictitious Religion C
Holy Trinity	Reject Trinity	Return of the Saviour
Non violence	No consumption of flesh	No consumption of alcohol
No Sacrament	Sacrament	Sacrament
Worship of one God (not denying existence of others)	Deliverance from evil	Matter does not exist
Personal sacrifice to benefit others	Confession	Worship of one God (not denying existence of others)
God by reason alone	God by faith alone	Non violence
Veneration of ancestors	Worship of one God (not denying existence of others)	Deliverance from evil
Procreation sinful act	Polygamy	Women subservient
Deliverance from evil	Non violence	Veneration of ancestors
Infant baptism	Personal sacrifice to benefit others	Baptism of adult believers only
Death not final	Veneration of ancestors	Personal sacrifice to benefit others

Table 2 Structured Doctrine Arrangement, with Universal Introduction

Fictitious Religion A	Fictitious Religion B	Fictitious Religion C
As B	Worship of one God (not denying existence of others) Non violence Personal sacrifice to benefit others Veneration of ancestors Deliverance from evil	As B
Holy Trinity	Reject Trinity	Return of the Saviour
Death not final	No consumption of flesh	No consumption of alcohol
No Sacrament	Sacrament	Sacrament
Procreation sinful act	Polygamy	Women subservient
God by reason alone	God by faith alone	Matter does not exist
Infant baptism	Confession	Baptism of adult believers only

This suggestion thus entails the following pattern:

1. All pupils follow a universal introductory syllabus comprising doctrines common to all the co-operating religions thus gaining a mutual understanding and tolerance towards each other that could last their lifetimes.

2. On conclusion, pupils are encouraged and guided to select responsibly for which religious (or non-religious) approach suits them best, or to continue with their family beliefs but see them in perspective against other creeds.

One can predict certain criticisms that will be aimed by the clergy, e.g. that this is a 'pick and mix' approach for the young to choose their own religion after the initial universal teaching at school. But why not? If clergy of the different denominations, either separately or together, were invited to address the young as part of this introduction, the young would eventually be in a position to make a mature choice. Is the present method of enforced indoctrination really superior? It is true that clergy would have to 'sell' their beliefs by being honest and clear in their presentations, but this would not hurt them or their causes, and the fears of losses in believers to other faiths should be offset by gains from these other faiths if their teachings are really worth while. If, on the other hand, the clergy claim that they have insufficient doctrines to offer such a universal introduction, is this really something to be proud of?

In short, the above proposal permits every religion to endure, but enables the young to be offered an independent and broad approach to all religions, and so enables them to make a mature selection rather than be dragooned into one only with no opportunity to refuse or to gain true knowledge of other areas of religious thought. Furthermore, it means growing up in a religious environment common with all their fellow pupils rather than being confined within separate boundaries laid down by their elders. If they then grow up to show greater tolerance to believers of other faiths, is this not a prize worth striving for?

Although the above ideas have not been presented as fluently as perhaps they might, the principle is clear enough, and as wide a discussion as possible is sought.

Matt put down his pen with some satisfaction. At last he was getting his thoughts sorted and laid out ready for constructive criticism. Andrew had proved to be a most kind and unselfish listener, and no doubt had realised that Matt had a great deal on his mind, and a need to unload it.

It was at this point that the telephone rang. "Hullo, Matt dear. I wonder if you'd mind picking me up from Sally's house."
"Of course, my love. Is everything all right?"
"Not quite. I'm afraid I coughed up some blood when we left the swimming pool just now, so I'll have to get it sorted."
'Oh No', thought Matt. 'Not again. Don't tell me we've got to go through all this grief again. Please, please, not again.' "Coming right away," he replied, through gritted teeth.

Then, pushing his manuscript into a desk drawer, he grabbed his car keys and made for Sally's house. Brenda's smile was a little forced, but she settled into the car, and they went round to the surgery to make an appointment. This led to a visit to the X-ray Department in the local hospital, and to an anxious wait for a result. The prognosis was not good. There was evidence of developed cancer in one lung, and a trace in the other, so immediate treatment was recommended with radiation and chemotherapy. The latter comprised continuous injection of drugs from an automatic machine rigged up on a wheeled stand, which meant that the patient had to wheel the stand around when moving in the hospital ward or going to the toilet. Periods of a few days of this treatment every few weeks were recommended. Brenda endured this cheerfully, and it resulted in some improvement. When the injection technique was superseded by tablets, the restrictions were greatly eased, and Brenda could then move about freely and follow a near normal lifestyle at home. They were also encouraged by the fact that Phyllis had been cured completely, and that advances were being made constantly in the treatment of cancers of all types.

* * * * *

A letter arrived from Hugh, the deputy head of the Engineering Department, to the effect that the Department had been invited to make a presentation regarding the current issue of the merging of some of the various engineering institutions, and that he, Hugh, had been tasked to prepare some preliminary notes. Knowing of Matt's sojourn in Australia, and his many years in the academic world, Hugh wondered whether Matt could find the time to jot something down to help in this regard. Matt felt flattered that he had been asked, and set to right away with the following results:

Dear Hugh,

Many thanks for inviting me to comment. I hope the following notes will prove useful:

<u>Notes on</u>

<u>Engineering Today</u>

"The Applied Science discipline of Engineering has had a chequered past, with applications both in peace and in war. The first signs of a primitive technology had begun to emerge several thousand years before Christ, with the appearance of burnt brick for building, asphalt for waterproofing, the smelting of copper, and the first fossil fuel age of the Babylonian empire. In more recent centuries, the engineer was recognised for the warlike operations of devising weapons and machinery for piercing defensive positions and wreaking human destruction, which were hardly conducive to sympathetic regard by the mass of the populace. In fact, it was only some 200 years ago that the peacetime value of the engineer became recognised sufficiently for a distinction to be drawn between military and civil engineering, and even more recently for civil engineering itself to split into the other parent branches of mechanical, electrical, and chemical, and then develop further into the many specialist branches available today, each with its own Institute, Institution, Society or other learned body.

This system served us well until recently when engineering has become such a vital issue that the government of the day must be made fully aware. However, with a multitude of specialist institutions in place, governments may be unsure of which one(s) to approach for advice. Being able to make a fresh start without too many constraints of history, the Australians solved this problem effectively by combining many of these specialist bodies into one unified Institution of Engineers, Australia (now known as Engineers Australia). Their governments thus have a single, recognisable and respected body to which they could, and do, turn for professional advice. In short, this Institution has political clout, and its voice does get heard in governmental circles.

Lately, similar moves have been recognised in the UK, and it is heartening to note, for example, that the Institute of Energy and the Institute of Petroleum have now combined to become the Energy Institute, using a single headquarters, and staff. The Mechanicals and Electricals have been in discussion over a similar process of merging, but so far without success. The Mechanicals have an aeronautical branch, although there is also a prestigious Royal Aeronautical Society dating back to 1866."

I would suggest, Hugh, that you make some reference to the 'umbrella' bodies that already exist, like the Royal Academy of Engineering, the Engineering Council, and the Association of Professional Engineers, commenting on their roles, their successes, and any failures. I think this would give a strong argument for merging, and I wish this move every success.

Thank you again for contacting me.

Sincerely

Matt

It was at this time when Matt was refuelling his *Honda Civic* diesel that another motorist nearby appeared to be in a somewhat flustered state. "Excuse me, chum," he ventured to Matt, "but I can't get this nozzle into my tank filler. What in hell's wrong? Any ideas?"

"Is your car petrol driven?" queried Matt.

"Well, it's the wife's actually, but yes, it is."

"It looks as though you've got a diesel nozzle in your hand, and they're designed to be too large to fit into a petrol filler pipe – for safety reasons."

"Oh my God, yes I see. I normally drive a diesel myself, but I just forgot I was driving the wife's today. Thanks very much, my friend. I'll do the job properly now."

After a pause to refill with petrol, the questioner asked. "Incidentally, what would have happened if I had been able to fill up with diesel, I wonder?"

"Hmm, think oy'l get a new tract ter with me Lotto win"

"It would have been pure grief," replied Matt. "Once the diesel fuel reached the engine, it would have started spark knocking severely, and punched holes through the piston crowns. It would mean a new engine, and also about a hundred pounds to have the tank emptied and cleaned. Mind you, had it been petrol put into a diesel tank, the engine might have just stopped dead without damage, but the fuel system could be wrecked through loss of lubricity. It's surprising but, according to the AA, there are over a hundred thousand misfuels of petrol into diesel vehicles every year, and they represent a loss of about four thousand tons of discarded fuel. Amazing, isn't it? But it's so easily done when you're used to using both fuels but are preoccupied at the time, as you've just proved." The motorist thanked him again, and drove off, shaking his head in relief.

* * * * *

Eventually, since Brenda seemed to have regained quite reasonable health, with the treatment having positive effects, they wondered whether they could contemplate another coach holiday. Their eyes fell on a particularly entrancing prospect of a week's journey from Land's End to John O'Groats, taking in a tour of the Lake District en route. Although the Norwich-based tour company warned that this holiday was not recommended for people with walking difficulties, Brenda was able to convince Matt, rightly, that she could manage a reasonable amount of walking, so Matt went ahead and booked.

As with most coach holidays in that part of the world, this one started at some early hour outside *John Lewis* at All Saints Green in Norwich. And, as usual, Matt and Brenda felt particularly relaxed when all the preparations, and the taxi ride, were completed, and they could sit back in the coach and switch off the moment it started. As a feeder coach, it threaded its way down the A11, round the M25 and west along the M4 to the Reading Services where they picked up the tour coach itself. Then it was the first long leg of the journey via the M5 and A30 to Newquay. They found the view from the hotel windows of the famous surfing beeches quite outstanding, with the wide sweeps of sand and continuous lines of breaking waves, and had plenty of time to enjoy it because their stay there was for two nights.

The following day saw their visit to Land's End, where all the holidaymakers had their photographs taken in turn at the famous signpost, each group with the distance shown from their home town. In Matt and Brenda's case, the photograph showed Norwich as 415 miles away. Things seemed to be very different from the last time Matt had visited the place. Gone were all the small houses with their lean-to workshops churning out holiday gifts – instead it seemed much more organised with restaurants, a visitor centre and children's play area, plus a helicopter and lifeboat on show. The party then returned to Newquay via the pretty little town of St Ives with its whitewashed houses located round two sandy bays. Brenda was particularly pleased to see it again since she had spent a holiday there with Norman some years before.

The second long leg of their journey took them for a two-night stay at Bassenthwaite in the Lake District. Once again, this was followed by a leisurely tour, this time through the exciting Kirkstone Pass, named after a large stone resembling a Scottish kirk located at the roadside. They enjoyed a 'wigwam' (teapee) at the 15th Century Kirkstone Pass Inn (with doors two-and-a-half inches thick!). Here they were told about the Herdwick sheep that are born black but turn grey within three years. Apparently, their wool is so coarse that they are able to survive in snowdrifts, gaining sustenance as well as protection, from their wool. One surprising fact was that there is only one lake in the so-called Lake District, and that is at Bassenthwaite – the remainder are 'Waters'.

Their next stops were at Ambleside and Grasmere, where Brenda was able to buy some wool for a tapestry she was planning to finish and give to Dulcie. At Keswick they visited the Derwent museum of the Cumberland Pencil Company and saw the 7-foot long pencil, the largest in the world. It appears that graphite (carbon, known locally as Wadd or Plumbago) was discovered at Borrowdale way back in the early 1500s. It was first used for making cannon-ball moulds by the Elizabethans, and taken to London by armed stagecoach for the purpose. However, the local farmers found it useful for marking their sheep, which led to the development of wood-casing. Stealing of the graphite and subsequent disposal led to the expression 'Black Market'. One exhibit of which Matt had been made well aware was the wartime subterfuge of incorporating escape maps and compasses in pencils of standard size which even contained sufficient lead to permit writing. Matt and his team had been issued with these, and also collar studs containing compasses, prior to their crossing the Channel during Operation Overlord in case they found themselves behind enemy lines.

The third long leg took them via Carlisle to crossing the Scottish border. The tour manager, himself of Scottish descent, was already wearing his tartan trews that day, whereas the driver donned a tam o'shanter, complete with ginger hair hanging from the back. The coach intercom was blaring with bagpipe music, and everyone went wild with excitement, feeling about ten feet tall. Definitely, one of life's outstandingly memorable moments. Then, after a short break at Gretna Green, it was a case of settling down to steady driving again, past Glasgow for a refreshment stop at Perth.

Matt had always been fascinated by the skill of glassblowers, and both he and Brenda were intrigued to witness molten glass being turned into highly attractive paperweights and many other articles at the Caithness Glass Visitor Centre. On reaching Inverness, the party had crossed Scotland from the West to East coasts, and they then followed the coastline to Latheron where they struck inland to Thurso on the north coast for their next two-night stay, passing many shaggy highland cattle on the way. One melancholy aspect of their journey was the sight of remains of abandoned cottages following the 'Clearances' of local farmers by the Duke of Sutherland – who still had a statue erected to him! Thurso has that grey-yellow granity look that typifies Scotland in the mind's eyes of southerners like Matt and Brenda, but any coldness in the appearance of the town was completely offset by the warmth and kindness of the members of staff, and in fact of all the assistants of the shops visited by the holidaymakers.

On the following day they were taken to Dunnet Head, the northernmost point of mainland Britain, where the wide aspect over the Pentland Firth and the Orkneys was much enjoyed by all. Then it was straight to John O'Groats. Apparently the name derives from a Dutchman, John de Groot who operated the ferry to the Orkneys in the 15th Century for a fee of four pence i.e. one groat, per journey. Because he had eight descendants, John built a house on an octagonal plan, and the present hotel built beside this site has an octagonal tower. Here the second of their commemorative photographs was taken, this time showing the distance from Norwich as 661 miles.

On the return journey to Thurso, they were fortunate to be able to visit the Castle of Mey, bought by Queen Elizabeth the Queen Mother in 1952, following the death of her husband. A guided tour of the castle included portraits of former owners, in particular the 14th Earl of Caithness who helped to create employment in the community by opening many flagstone quarries – he was responsible for introducing the first steam-driven car into the country in the 1870s. The fruit and vegetable gardens were also pleasant to visit, and shelter from the wind was evident in the shape of a high protective wall and copious thick hedges. The Queen Mum's own chair was then indicated in the local church.

A melancholy note crept in then as the coach turned south the next day to head for home, this time staying one night at a hotel in Teesside, although Matt was delighted to see the massive Forth Rail Bridge, and even photograph it through the coach window. However, the huge statue of the Angel of the North left him cold. He just could not see the point of erecting tens of tons of steel to link the North with the supernatural. 'Are we to have Angels of the South, West and East,' he mused. 'If not, why not?' He could not help feeling that the metal could have been used in a much more useful way. The Friendship Bridge in Norwich, for example, combines emotional overtones with a useful purpose in permitting people to cross the river from either side, and at 85 tonnes probably uses far less material. But that was typical of his thinking – i.e. Nature rather than Supernature.

Feelings are always a little flat at the end of a holiday, but Matt and Brenda said goodbye cheerfully enough to the tour manager and driver at Newport Pagnell services, and joined the feeder coach for Norwich. They passed within a mile or two of Brickhill but could not, of course, stop by to visit. They quickly readjusted to normal routine on reaching home, but Matt could not completely erase the nagging thought that this might be the last long holiday that Brenda and he may enjoy together, although he did his best. After all, she had stood the journey very well, and the periods of drug transfusion in hospital were now replaced by taking tablets at home.

Matt and friends were saddened when the time came for the three remaining Concordes to circle Heathrow and make their final landings before being dismantled and sent in pieces to museums. The familiar engine roar could be heard clearly in Norwich, and seemed to act as a requiem. This magnificent aircraft had proved to be a marvel of both technology and Anglo-French co-operation, with only one major disaster caused by debris on the runway rather than the aircraft itself. However, Matt had to bite the bullet and admit that, if it did not pay its way, there was a strong argument to cancel it.

It had now become clear that Brenda's health was beginning to deteriorate. The X-rays showed no further spread of the cancer, but her energy level was lower, and so Matt borrowed a wheelchair from the Red Cross. Also worrying was the fact that Matt's own health was suffering, due largely to the worry over Brenda, and his almost full-time involvement in housework, and his weight was showing a steady decline. The family rallied magnificently but, of course, they had their own lives to run, and Matt could not let them overdo their kindness. Matt and Brenda realised that a wheelchair may well be a permanent requirement, so they bought one for themselves. There were also some money worries with a mislaid cash card, all adding to Matt's nervous stress.

There had been a concern niggling at the back of his mind that the price of houses was rising so fast that the value of the *High Firs* property might exceed the accepted barrier, in which case his dependants would become victims of the notorious 40% inheritance tax. He passed this query over to his accountant and, to his horror, found that Steve and Avril would be faced with a bill of several thousand pounds payable on his demise. He immediately arranged the transfer of *High Firs* to Steve, Avril and himself in equal thirds and hoped that, for this and many other reasons, he would live out the seven years and rid them all of this incubus. It annoyed him intensely to feel that the various Chancellors of the Exchequer had eyes only for taxing people as much as they could get away with, and had no idea of encouraging peoples' instincts to save. After all, everything that had been put into the house in the way of goods and workmanship had already been subject to tax, and it's only common sense to reward people who have conscientiously cared for their property, rather than spending it on luxuries and holidays, by permitting them to hand these benefits over to their families. The family, in this case, had not just sat around waiting to inherit – both Steve and Avril had put in hours of work repairing and maintaining *High Firs*, so surely they were entitled to inherit, with a modest inheritance tax only based on a high barrier and a very low tax rate on the excess.

Matt had always been angry that the Inland Revenue seemed to have no idea at all of making the tax system simple and intelligible to anyone with any reasonable sense. The Australians had done it, so why couldn't we? Matt and his academic colleagues were spending their careers in making complex things simple, and this sort of expertise was sorely needed by the tax authorities. Matt was certain that, with many hours of studying, he could understand the British system, but what a waste of time this would be! Much better to spend it on his own subject. And in any case, the system would no doubt be changed every year. In Matt's view, these politicians had no real understanding of human nature, and were concerned only with the bottom line of their accounts appearing in credit. In short, they were no better than bean counters!

Matt's aversion to things political had become well known to his colleagues, and he had been asked, somewhat jocularly, whether he thought that the country should be run by applied scientists. "No way," was his reply. "It's a very healthy thing for applied scientists to submit their views to a higher authority for critical assessment. But, that said, the higher authority must listen, and must be wise enough to assess the merits of the argument, and judge accordingly. Somehow, the strength of the applied scientists' input must be increased." This is what he had in mind when submitting his notes on institution mergers to his colleague, Hugh. Matt considered that the English political system was the best in the world – it was just the people running it who cruelled it. In any case, those involved in science are not saints or knights in shining armour. There had been some terrible scandals in the past. But the point was that it is characteristic of the scientist to report honestly, whereas this cannot truthfully be said in the case of the politician who relied on spin doctors to massage the facts for his own benefit.

It was all this mental upheaval that led to a very unfortunate development in that Matt, not clearly thinking what he was doing, lifted Brenda in her wheelchair up the kerb in a rather clumsy manner. Noticing some discomfort below later, he visited his GP who announced that he had a fairly extensive hernia!

Matt felt that this was the breaking point. Although he had not yet accepted that Brenda's illness was terminal, that possibility did lurk at the back of his mind, and he wanted her to feel that she could rely on him always being with her to care for her to the very end. But now, through his own stupidity, he had compromised his ability to do this, and this realisation resulted in the tears flowing unchecked down his cheeks, even in the doctor's surgery.

"How long would it take to get it fixed, Doctor?" he queried anxiously.
"Probably about eight months on the present waiting list." "Could I go privately then, and get it done right away?" "Well, you could, but might it be better if you spend all your time caring for your wife now, and having this repair done later?"
The implication was dire, and obvious.

Matt accepted this advice, but his spirits were at rock bottom. This gloom was deepened when, at Brenda's next meeting with the specialist, it was pointed out that the cancer was spreading, and that an appointment with a hospice was recommended. So nothing more could be done for Brenda, and Matt struggled through nursing her as best he could, trying not to show that he was crying inside. For her part, Brenda was remarkably stoic, never once shedding a tear, although she did express regret that she would miss the great joy of seeing her grandchildren grow up into adulthood. She tried to occupy her time by advising Matt on the housework, and continuing with her tapestry work that she wanted to finish and give to Dulcie. Also, whenever Matt could be persuaded to drag himself away from the kitchen chores, she wanted him to sit with her on the couch and comfort her by cuddling her as they always used to do. A particularly poignant moment came when Brenda found that she didn't even have the strength to push the needle into the tapestry!

Matt's health continued to deteriorate, with nothing he could do to stop his weight loss. He also developed some phlebitis in one leg, and was troubled with piles, acute constipation, and benign prostatic hyperplasia leading to frequent visits to the bathroom at night. Eventually it became clear that Brenda could not last much longer, and both Dulcie and Margaret moved in to help care for her, while Matt was moved to the spare bedroom. Then, one night while they sat round her bed waiting to see how they could help her, her breathing suddenly stopped, started again for one last breath, and then stopped for ever.

Rather than breaking down, Matt felt singularly calm, and was able to organise the doctor and the undertakers with controlled effectiveness, although he felt, once again, that his insides were being held by a vice. He handled the official procedure of registering the death and dealt with the necessary paperwork as if in some kind of dream, or some parallel life, and then began to realise that he was approaching a nervous breakdown, just as he had seen some of his colleagues when they were under great stress. He did his best to answer all the kind friends who had sent messages of condolence, but eventually had to give up and hand over to the family.

Mercifully, Geoffrey took over the organisation of the funeral, and Matt attended the church service, because he knew that this is what Brenda would have wanted, but he just could not face attending the crematorium afterwards. He was much heartened when Wendy, sensing his desperate sorrow, slipped her arm in his as he left the church.

They had arranged beforehand that, should this eventuality arise, Matt would hand over the bungalow to Brenda's family and return to his own family home at Brickhill. Steve, Margaret and Avril helped him pack, organise a removal van, and transport him back to Brickhill to be nursed back to health. It was then that Matt suffered a further blow to his equanimity, and his temper, when he learnt that, even though he was a joint owner of *High Firs*, his family home, the law would not permit him to live there! This catalysed him from his depression, and he made immediate contact with his solicitor who was able to prepare a suitable document to authorise his continuous residence there. Matt was, not surprisingly, extremely relieved, but his opinion of politicians, and politics in general, had now become unprintable! 'Perhaps applied scientists *should* take over, after all. These xxxxxxs just have no common sense whatsoever.' (Fortunately, he did not learn until much later that, under certain circumstances, inheritance tax may have to be paid *twice*! Had he done so, he may well have blown a fuse on the spot.) As a red rag is to a bull, so are politics to Matt, which is a pity because he might have had a useful part to play in running the country on an honest, common sense basis, with not a spin doctor in sight, nor any mark where one had been.

As soon as he could, Matt organised surgery for his hernia repair. This was expensive but, to Matt, worth every penny. The surgery went well but, due to the extent of the hernia, there were some resultant complications in the form of internal bleeding and secretion of serum. The surgeon, who was professionally kindly and encouraging, assured Matt that these problems would sort themselves out in due course, but Matt still felt that his life, or working life at least, was now over, and that he might as well clear out all his notes and books. However, the surgeon's predictions proved correct, and the tamsulosin prescribed by the GP controlled the prostate problem, so Matt started to pick up again with his abstract service. He also had another invitation to lecture at the university, but Professor Taylor was kind enough to suggest that, initially, he might like to tackle one hour only, rather than his usual two or three.

Naturally, Matt recalled how wretched he had felt when he lost Joyce but, of course, he was younger then, and his physical health at that time was sound. This was the second grievous blow he had suffered, and he found that it did not get any easier the second time around. But gradually, following his recovery from the surgery, he found his health steadily improving almost back to how he had been a year or two earlier, and he discovered two ways of helping himself to keep swimming in the sea of despair rather than sinking into it. The first was to work, physically and mentally, in order to occupy his mind rather than linger over those lovely lost years. The second was to compare his condition now, with what it had been when in the depth of depression, rather than with when he was at the peak of happiness with a healthy, loving wife.

Having recovered from the loss of Brenda, Matt hoped he might eventually deserve the epitaph, 'Fate managed twice to break his heart, but failed to break his spirit!'

Another sad event had to be faced with the disposal of his beloved boat. *Pearl* was duly put on the market and, following the tender loving care that Matt and the family had expended on her, a quite reasonable price was achieved. Matt had loved those hours on the water, but he realized that that pleasure, too, had gone for ever.

Slowly, a new pattern of life emerged for Matt as he and Steve's family settled in to the changed arrangement at Brickhill. Matt was thoroughly relieved to be fit enough to drive again, and to lift reasonably heavy weights, now that the hernia had been repaired with a strengthening plastic mesh. He knew he had so much to be grateful for: a loving family, a comfortable home, an interesting occupation both at home and at the nearby university, and so many good friends. Even the two cats had deigned to recognise his permanent presence, and Penny had graciously adopted his duvet for her nightly periods of oblivion.

And yet. And yet. If only he had a loving partner of his very own. Would this yearning never cease, he wondered.

Chapter 22 – Aftermath (2004 – 05)

Matt finished his coffee, and added the mug to the pile awaiting the next wash-up session. He sat gazing out past Ozzy and the photographs of Joyce and Brenda at the varied vegetation in their 'nature' garden. The daffodils were providing a golden carpet gently bobbing in the breeze, and the squirrels were having a high old time winkling out the peanuts from the bird feeder. His gaze gradually lapsed into the middle distance, and he found himself wondering about the past, the present, and the possible future.

He'd had a stable family background, thanks to his parents, sister and twin brother, and some good friends to grow up with. Looking back to his young days, it was all so carefree, with their uncertainties about what was happening and how to respond offset by inextinguishable confidence that they all had so much to look forward to as they grew up to adulthood in such an exciting time. Those were the heady days of Croydon serving as London Airport; of *Imperial Airways* blazing routes across the globe; of outstanding aviation personalities like Amy Johnson and Jim Mollison, and of Scott and Black flying from Mildenhall to Australia in the de Havilland Comet 'Grosvenor House'; of the first flight over Everest; of the new breed of flying boats of C class type with classical names like 'Canopus' and 'Cato', and of the impressive but ill-fated R101 airship based at nearby Cardington, and the Hindenburg from Germany. He recalled how his bedroom walls had been plastered with pictures of these craft, and of the far-and-away places to which they had travelled. It was only later that he had managed to find space to squeeze in a picture of Jessie Matthews! Mike also had a bedroom picture gallery, but the ratio of pin-ups to aircraft was somewhat higher.

Matt's thoughts then roamed over his degree course, studying in London during the early years of the bombing, not knowing whether the university would still be there the next day – steadily learning how to learn, and making good progress in each subject, except perhaps mathematics. Algebra was fine – all so logical – and geometry came to him with little effort. He recalled how the lecturer had produced a set of three diagrams – plan, side and front elevations – of a complex system of pipe-work, and how that he, Matt, had been selected to combine the three into an isometric three-dimensional view. He also remembered with quiet pride how he had managed to solve it in record time – because he just had that sort of brain. But pure mathematics had tested him to his limit. He could handle ordinary arithmetic, and even calculus after a struggle, but when they reached complex numbers, Matt knew that he was beaten.

When told that a quantity 'x' could be written as

$$x = a + i b$$

where 'a' is a real number and 'b' is imaginary, he could see danger approaching. When told that term 'i', the operator, is equal to the square root of minus 1, Matt recognised that he had met his Waterloo. He had always understood the following relationships:

$$(+1) \times (+1) = (+1), \text{ hence square root of } (+1) = (+1)$$
$$(-1) \times (-1) = (+1), \text{ and, square root of } (+1) = (-1)$$
$$\text{also } (-1) \times (+1) = (-1)$$

So what on Earth was the square root of (-1)? His fellow students seemed to absorb the concept without difficulty, and he was told that electrical engineers used it when dealing with alternating current, except they call the operator 'j' rather than 'i'. It bothered him a bit until he realised that different people possess different skills and, in fact, the brightest mathematician in the group had some difficulty when attempting the three-dimensional pipe-work exercise. Matt was just short of that type of imagination.

Gaining the first degree was very exciting, considering the wartime conditions under which they had all had to study. It was rumoured that the authorities had offered to lower the pass rate in view of the situation in which the students found themselves, but the unanimous cry from the students was 'do not lower standards, otherwise our wartime degrees will be treated with contempt in future'.

Then there was meeting up with Joyce – the prettiest girl he had ever known; the traumatic loss of Mike just when he was in the flower of his young manhood with everything to live for; the blessed relief from heartache by marriage to Joyce, and the consummation after years of self control. Was it all worth it? he asked himself. There were times when he longed for a woman to enjoy and ease the naturally growing pressure within him. But if he had yielded, he reckoned that the temporary relief would have been offset by other, and possibly permanent, problems, judging by the experience of others. He was convinced that the bond created between Joyce and himself was as near perfect as it could have been. When various topics arose in company, their eyes had met for only a second or two to convey the deepest of understanding between them. There was rock solid confidence and trust between them, which passed the tests of stress with flying colours.

Then there were the children, and how exciting it was to see them arrive and develop into individual beings with traces of both parents in their looks and personalities. Thankfully they were both physically and psychologically fit, with none of the traumatic problems that some parents had to suffer. Both had gone on to make careers for themselves, then to marry and produce fine grandchildren.

But why, oh why, did Joyce have to be taken from them? She was a thoroughly good person; a wonderful mother and loving wife. And yet, a useful member of society like her was lost to us all. Matt recollected how desolate he was to lose her, and how his sanity was saved by the response from Brenda. In contrast to Joyce's chronic health problem, Brenda had proved a very robust person, so they both had the pleasure of enjoying a second chance which had proved intensely rewarding. And now she, the most handsome woman he had ever met, had also had to go. Why, for heaven's sake, why? There were plenty of women around, older and with nothing like the qualities of Brenda, who were tired of living and quite prepared to go. Why weren't they taken instead of such a charming and good-living Christian who had so much to live for with her loving family and doting husband who ached with love for her? Where's the sense of it all?

But sense did start to creep back into Matt's feelings. We are brought up to realise that our tenancy of this Earth is only temporary, hence we should be grateful and enjoy life while we have it. It does perhaps seem strange that we are born to grow up, procreate, and then steadily fade away while our progeny go through exactly the same routine, ad infinitum. But it's the other things we do during our lifetimes that also matter, particularly if we are minded to develop and use our talents to help others and to create something rather than subjugate and destroy. Matt was never overly impressed by high quality of clothing or personal adornment, for example. To him, what really mattered was what knowledge was assembled in one's brain, and skills in one's hands, as well as how these were used.

Matt felt he had climbed as high up the tree of knowledge as he was capable, and was now feeling rather tired. Fortunately his enthusiasm for his subject of fuel technology still endured, and he was so grateful that the professional institute in London was prepared to accept his technical abstracts, and also that the university still invited him to lecture. Both these activities served two purposes, as far as he was concerned. First, they gave him an incentive to keep up-to-date with the exciting developments in alternative fuels, hybrid vehicles, fuel cells, solar energy and the like. But more fundamentally, they kept his mind from lingering over those lovely past years with Joyce, then with Brenda, and so saved him from lapsing into deep depression, and finishing up as a sobbing heap, useless to family and friends.

He had been so lucky in his love life that he reckoned he should be doubly grateful and shut up moaning, in that he was being cared for by a loving family, in his own home, and now enjoying relatively good health. And yet. And yet. To be honest to himself, Matt still ached for feminine companionship, to hear again a woman's voice, and to see again the form of a woman as she weaved in and out of his daily life. Sex was not a major driver any more – it was companionship, understanding, sympathy and the sort of friendship that excuses one's mistakes, and endures regardless. A female friend to enjoy shared company. But, he reasoned, what woman would look twice at a bald-headed old wrinkly like him? Best to accept the inevitable as cheerfully as possible, and not spoil the lives of his family and friends.

So Matt took out his lecture notes again, blew the dust off them, and started to revise.

Matt had volunteered to do his share of the weekly shopping, armed with a well-prepared list. Routinely he would go to the food centre at Central Milton Keynes, but on this occasion he travelled to nearby Woburn Sands for the groceries, now that there was a spacious car park built behind the High Street. An elderly lady customer had just paid her bill and turned to leave the shop when her bag split, disgorging all her purchases over the floor. She was clearly deeply upset by this accident, and very close to tears. The shop staff and several customers offered help, but Matt took charge, reloaded her goods into another bag, and then guided her to his car where he invited her to sit to recover. He drove her to her home and saw her indoors, where an infirm bedridden husband thanked him profusely for his kindness. Matt enquired about their future plans for shopping etc, but they assured him that their daughter was on her way to stay with them, and that the lady was trying to help by collecting the groceries beforehand. With that, after a chat with the husband, who seemed to be desperate for male company, Matt wished them well and returned home.

As a matter of routine to maintain his health, both physical and mental, Matt undertook a daily walk in the local woods of the Woburn Estate, often calling at the Re-enTree to try to feel the contact with Joyce and with Brenda. On this particular day, as he was returning along a narrow path through the fir trees, he casually noticed a lady with her golden retriever proceeding in the same direction on the main parallel bridleway a few feet higher. The dog was enjoying his treat of being able to chase the ball thrown for him at regular intervals. Eventually, the ball rolled down towards Matt's path, and buried itself amongst a pile of branches. The dog came to a halt with his ears cocked, puzzled as to what to do next. Matt could see the ball below all the boskage, and bent down to reach through and rescue it. What he should have done then was to throw the ball out so that the dog could retrieve it and return to his mistress without any accidental biting from the excited animal. But Matt couldn't really care too much about any such hazard any more and so, gently holding the ball at two points, he offered it to the dog with encouraging words like, "Come on, boy. Take the ball from me, there's a good lad." The dog looked at him quizzically for a second, then did exactly as instructed, and returned happily to his mistress.

Matt thought no more about it, but plodded on in the general direction of home. As it happened, his path rose slightly to join the bridleway, onto which he emerged just as the lady and dog duo arrived as well. "Thank you," said the lady, "that was kind of you to rescue Bernard's ball for him. I'm so glad he behaved himself when you gave it to him."
"My pleasure," rejoined Matt, "I've always had an affection for animals, and they seem to accept me. I remember my father's old joke that a nearby dog must have liked him because when it bit him it didn't spit it out!" Matt smiled at the reminiscence.

The lady smiled rather wanly. She was somewhat younger than Matt (but isn't everybody, he would have said), and clearly had a cultured background. Like Brenda, and Joyce before her, she had retained her looks despite the passing of the years.

Matt thought, 'I must watch myself – looks a rather high-class type, and might not take kindly to any robust humour. Better change tack.' "I often come for walks in these woods," he said, before adding, "Wonderful for calming the soul, and such a relief to find some peace now and again. Incidentally, I live along the road there – have done for many, many years, except for one spell in Australia and another in Norwich. Must say it's good to be back. Hope you don't mind my talking to you like this. After all, I'm just an unknown man walking the woods."

"Not quite," came the rather surprising reply. "I saw you in Woburn Sands the other day when you were so kind to that unfortunate lady who had the accident with her shopping bag. That was a very neighbourly thing to do."
"It was the least I could do for someone who was clearly in distress."

Matt then went on to recount his giving transport to the lady's home, and the imminent arrival of her daughter. He went on to explain that the bedridden husband was only too anxious to have someone to talk to, so Matt had stayed until the daughter arrived. "It all helped me, too, because it gave me a chance to do something useful, and stop moaning about the past."

Before he realised it, Matt had given a thumbnail sketch of his marriages and losses, and how he was coping with the situation by a routine of concentrated work and regular exercise. He found it very easy to converse, and before either of them knew it, they had arrived at the lady's home. Sensing an imminent supper, Bernard bounded in through the gate which, Matt noticed, was not held shut properly because the screws holding the catch had become loosened.

"May I put that right?" asked Matt, and before receiving a reply, took his famous screwdriver out of his top pocket and promptly twisted the screws well home.

"Well, thank you very much," ventured the lady. "Incidentally, my name is Pamela, and as you know, the dog here is Bernard."

"Thank you for that. My name is Matthew, but I'm generally known as Matt by all and sundry. If I should see you again in the woods I'll know how to address both of you properly. So goodbye for now."

* * * * *

Several days later, a letter arrived from Andrew.

> My Dear Matt,
>
> I have only just heard from former colleagues of your sad loss of Brenda. You have already experienced the loss of one loving wife, so this must have been doubly traumatic for you. I am so deeply sorry, as I know how much she meant to you.
>
> They tell me that you are coping well and busying yourself in work, and I congratulate you on this in the hope that it will comfort you sufficiently while time helps to ease the pain.
>
> As you may have heard, I took early retirement from the university, and now live with Ellen in Shropshire close to our children. Also, our grandchildren are not far away so we keep in touch regularly. My reason for leaving centred on this wretched arthritis, but the doctors have managed to stabilise me well now, and I am enjoying life once again. Had I completed my full tenure at the university, I would, of course, have thrown

the traditional retirement party, at which you would have been an honoured guest, but I just wasn't well enough then.

I trust your health remains as good as possible, and would be delighted to hear from you if you can find a moment.

With kindest regards,

Andrew

After some deliberation, Matt prepared the following reply:

Dear Andrew,

It was good to hear from you, and I thank you for your kind thoughts. I am so sorry to hear of you indisposition, but glad that you are comfortable again now.

You are right in that I find solace in work rather than in alcohol, tobacco, drugs – or even religion! Regarding that, I managed to focus my thoughts on a positive proposal to try and improve relationships between different religions, and felt quite content about it since this idea has been floating about in my mind for many years. However, recently I have become quite despondent about its future possibilities. I have been watching on TV some of these splendid programmes of those presenters travelling the world exploring other cultures and their treasures, and it is obvious that religion has such a vice-like grip on the lives of so many people that there is absolutely no chance whatever of their religious leaders even contemplating the idea I've put forward for an international introduction common to all religions, let alone giving young people the freedom to make their own choice afterwards. So, apart from having the satisfaction of clarifying my own ideas, I think I've been wasting my time. I suppose it has served me well, but that is all.

I am enclosing a copy of my proposal, because you've always been so kind in listening to me when I've been sounding off with my thoughts. I just feel too tired to argue about it with anyone, and am content to quietly concern myself with things that lie in my own area of experience, like the technical abstracts, and the lecture courses that the university continues to invite me to present. I'm sure that the university has no idea just how kind it is being towards me, and how deeply grateful I am for the opportunity to update the short course delegates on the fascinating developments in the fuels world.

I hope very much that you enjoy your retirement, early though it is, and am sure that, like all of us, you will find yourself so busy that you'll wonder how you ever found the time to go to work!

Please convey my best wishes to Ellen and your family.

Sincerely,

Matt

After a few days, a reply arrived from Andrew.

My Dear Matt,

Many thanks indeed for your kind thoughts, and your proposal re religious teaching. It may surprise you (or not) to learn that I am taking up lay reading, because I feel that the Christian religion is so well worth disseminating. However, I am profoundly excited by your idea of a common introduction that is international and, like you, I feel that if it could be implemented it could not fail to ease tension between different creeds.

However, I quite understand how despondent you feel when we encounter 'iron curtains' between religions, but I give you my word that, although based on Christianity, I will make it my major objective to encourage the mutual understanding and tolerance that you advocate.

You may laugh at this but, quite frankly, I think that your proposal is so vitally important that this is the main reason you have been put on this Earth, and I therefore humbly suggest that you should now feel able to retire from this particular area of verbal battle, and leave me to pick up the baton and continue with the good work that you have started. You no doubt agree with me that it will take many generations for this kind of wisdom to take root and flourish, but each young generation seems to show an increasing ability to question the status quo, so I am convinced that flourish it will. I only wish we could both be around to see it happen.

Please keep in touch during the years we have left to us. We share, I know, a mutual respect and, in view of our differences in outlook, represent a model of tolerance.

My good wishes go with you always.

God bless.

Andrew

Matt was so impressed and touched with this response that he felt the need to reply once more.

My Dear Andrew,

Thank you so much for your kind offer to continue the crusade for us both. We have our differences, of course, but much more important are the things we have in common. It's just that we approach them from different angles.

For instance, we both respect honesty and decency, and both support communication and communion between peoples, which I think we can summarise as:

We hope all languages endure,
but recommend English as the most suited to international use.

and

We hope all religions endure,
but recommend a common initial teaching and subsequent responsible choice.

We also agree that life has multiple benefits to offer, like music, art, literature, fitness, falling in love, parenting, food, interesting work and holidays, so it seems tragic that we have to make war instead of joining forces to fight disease, poverty and ignorance.

I consider it a privilege as well as a pleasure to know you, Andrew, and wish you and yours every happiness.

Kindest regards,

Matt

* * * * *

The early months of the year had seen the growing airliner rivalry between Boeing with its 777 and Airbus with its massive A380. As far as Matt could see, there was room for both vehicles, since the former had the advantage of range (10,500 as against 8,000 miles) and the latter with passenger capacity (555 to 840 in comparison with 301). However, he wondered how the airports would cope with handling so many passengers at the same time, particularly if both aircraft arrived together! Despite his years in aviation, Matt was still intrigued to find that the A380 would carry a fuel load of nearly 300 tons! How the Wright Brothers would have been impressed.

On one of his daily walks in the woods, Matt noted the Pamela-Bernard pair strolling along some way away, too far to exchange greetings without shouting, so Matt kept quiet. However, Bernard took control of the situation and, recognising a human being who had been kind to him, stroking him round the shoulders and muzzle in a firm but gentle way that he enjoyed, and who smelt good, he came bounding up to greet Matt with all the appearance of a smile on his face. There was no need for pretence now of not noticing each other, so they approached and, once again, exchanged pleasantries. Yes, apparently the gate catch was still holding on, although Matt suggested that perhaps some longer screws would be appropriate and, if acceptable, he could supply from his extensive workshop collection and then fit these – a job that would take no more than a few minutes. Pam accepted gratefully, as long as he was sure that it would be no trouble. "Not at all. After all, this is what neighbours are for, isn't it?"

Matt went on to explain that he dabbled frequently in DIY projects, but that he did have the sense to know when a job was beyond him, and thus when to call in the experts. For that reason, he had built up some good connections with skilled tradesmen in the village, and would be happy to offer their addresses should she ever need them.

"Well, there is one item that needs looking at. You see, my husband put up a shelf for my cookery books, but he didn't have much idea of going about it, and it's now sagging at one end. I would appreciate some advice." 'Oh dear,' thought Matt. 'This is getting complicated. A husband! I wonder where he is now.'

"I should have said my former husband," explained Pamela. "He was very good at making money – and friends. So much so that he befriended his secretary and went off overseas with her. Henry and I are divorced now, but I have been left with the house, and Bernard." All this was said more in sorrow than anger, but Matt sensed the hurt, and the feeling of rejection that must have been aroused. He couldn't quite decide whether it was worse to lose a loved one in death, or an unloved one in divorce, so, apart from some muttered condolences, he kept quiet.

Their friendship, though recent, had become robust enough for Pamela to invite Matt in to advise re the kitchen shelf. Seeing that the screws had been entered directly into the plaster without plugs of any kind, it was instantly clear to Matt that, although Henry had been an expert in the worlds of business and lonely secretaries, he could not have had much idea on which end of a screwdriver is which. Matt explained that the job should be done properly with wall plugs, of which he had dozens at home but, in the meantime, he could just put the screws back in sufficiently to hold the shelf as long as it was not loaded with books.

"I'm afraid my pocket screwdriver is not man enough for this, but I could pop back home for a twelve-inch version that should do the job."
"Well, thank you. Perhaps one of the tools in the garage would be suitable. Henry always had a box of them there, although he didn't often use them. It's through that door if you care to have a look."

Matt walked into the garage and there, to his amazed delight, was a magnificent *Jaguar XJ6* – in black – a car that Matt had always considered to be the most beautiful machine that had ever been built. The designers had got the line of the bonnet, cab and boot just right, and Matt always drooled whenever he saw one. These was no possibility of his ever having one, of course, and even if he did it would not fit in his garage, but there was no harm in looking and longing. He had had the privilege of driving Peter's *Bentley*, and also an opportunity to take a *Toyota Prius* hybrid car round Norwich, but he had never had a chance to drive a *Jaguar*.

Returning to the kitchen with a long screwdriver that was somewhat battered and covered with dried-on paint, Matt congratulated Pamela on her choice of vehicle. "Oh, it's not mine. It's Henry's, and he's sending a man along shortly to take it away. Which reminds me, I must think about getting myself a new car now. Something small and economical. I don't like driving, but we do need transport, don't we, when living in the country?"
"We certainly do. When you're looking at new cars, you might care to consider a *Toyota Yaris*, or a *Honda Jazz*. They're both small and manoeuvrable, and fuel consumption is low. You might even think of going for a diesel, as I have. They're so much quieter now than they used to be, and the particulate problem is very largely eased."

"Thank you for that. I would much appreciate some technical advice when the time comes."

Matt tightened up the screws as best he could, and promised to look in with some proper plugs as soon as it was convenient to her. Before the agreed date for the repair, however, an agitated phone call came through from Pamela to the effect that Bernard had been the victim of a hit-and-run accident, and was lying in the front garden in some distress. "I'll bring my hatchback round right away," responded Matt, "and we'll get him down to the village vet." Together they lifted the blanket on which Bernard was lying, and eased him into the back of Matt's car. Pamela was clearly distressed. She had already lost a husband, and seemed now to be losing her only faithful companion.

The vet completed his inspection, and reported to the effect that the impact had pushed many of Bernard's vital organs up towards his shoulders, compressing his lungs in a dangerous way. However, he considered that major surgery could just put things right, but it would be major, and the cost would approach something of the order of a thousand pounds. Did madam wish him to proceed?
"Yes, please, without any doubt. He means so much to me."
"Right. Then leave him with us, and we'll start as soon as we can prepare our operating room. We'll ring you as soon as we have any news."

Matt offered to take Pamela into a local tea room for a coffee, but Pamela replied that she would prefer to be at home next to her telephone, and would Matt care to join her there for a while? And so the two of them took to the kitchen and talked fitfully about the situation while Matt effected the repair to the shelving. Matt tried hard to keep the conversation going in order to take her mind off the present trauma, but it was not easy.

Eventually, the phone rang to give a message that the surgery was over, and that Bernard was now sleeping off the anaesthetic, having apparently come through successfully. He would be kept at the vet's for a day or two to ensure that his progress continued uninterrupted. Matt felt almost as relieved as Pamela must have done, and, having accepted her sincere thanks for his instant help at a critical time, he drove back home and had an early night.

It was when Matt helped to bring Bernard home that he ventured to suggest something that had blossomed in his mind. "Look, Pamela, you say you don't enjoy driving very much. I wonder if you would care to come out with me to some local places of interest. I'm a life member of the National Trust, and there are a number of lovely old properties around here that could be worth a visit. Then there's the National Garden Scheme, with some beautiful gardens to see for a modest donation to charity. And of course there's also Bletchley Park with the fascinating enigma story, and many museums and theatres…"

Matt stopped short there. He mustn't gabble on too much in his enthusiasm or it might frighten her off. "Well, I don't know. I guess I'll have to think about it. But thank you so much for the suggestion."
"But you will think about it, won't you?"
"Yes, I promise."
"And will you phone me, whether you agree or not?"
"Yes. Whether I agree or not."

Matt drove home again, feeling that he had got as much out of the day as he could reasonably expect, but with the faintest of hopes for future companionship.

Chapter 23 – Phone (2005 – ?)

Nothing much seemed to happen over the next few days, so eventually Matt stoically took out his lecture notes once more, blew the dust off them, and started to revise.

"Dad! Dad!" Margaret's voice came rocketing up the stairs.

"There's someone on the phone for you. Name of Pamela."

Appendix 1
Preparing for Science-based Examinations Success

What this appendix **IS NOT**: An easy option for the lazy.

What this appendix **IS**
A structured and disciplined approach to condense into digestible components the mass of material presented during an examination course.

Introduction

Each student must, of necessity, choose his or her own style of learning, but the following 'Summary' method is offered as a much-tested option that has resulted in outstanding examination success.

No claim for basic novelty is made here, but this approach is not always described in depth elsewhere, and hence may be overlooked during the stressful days of study. However, when followed it serves both to identify the key data, and to fix them in the memory as the course unfolds. This in turn also makes feasible an effective last-minute revision, eliminating the oh-so-common pitfall of promising oneself a final read-through on the examination eve, only to be overwhelmed by a bulging file of unedited lecture notes. A candidate so hampered enters the examination hall hoping against hope that certain questions *will not* appear in the paper. In contrast, our efficiently-prepared candidate enters hoping that certain questions *will* appear, armed with a quiet confidence based on sound revision. This difference is the vital ingredient for success.

The 'Summary' Method

1. After each lesson, prepare a summary (preferably in less than a single sheet) listing the major equations, formulae, definitions etc, with even a sketch of a key diagram if appropriate. Aim to be as *concise* (brief but not omitting key items) as possible. Do not postpone this action too long after the lesson otherwise some important spoken but unrecorded items may be forgotten.

2. File this summary sheet in a Master Summary folder for the given subject.

3. Make the summaries continuous for the given topic, but start a different topic on a fresh sheet. As the sheets accumulate, be prepared to rewrite them as necessary. Such a revision not only gives opportunities for corrections and improvements, but *helps to cement the material in the memory*.

4. Tackle some simple questions relevant to the topic and, once certain of accuracy, file them immediately after the relevant summary sheets.

5. When appropriate, tackle some past examination questions relevant to the topic and, once certain of accuracy, file them immediately after the relevant simple questions.

6. On the few days prior to the examination, pack the lecture notes away and concentrate attention on the Master Summary folder. Since this will be a relatively slim volume giving instant access to correct solutions of examination questions, this task will not be as forbidding as tackling a thick file of original class notes. By this time, the main data and their applications will already be sitting comfortably within the memory circuits, hence the final read through will re-charge them without stress.

Ideally, this method should be applied to each lesson in each examination subject. However, do not despair if a few gaps arise through pressure of other commitments, as long as the majority of the source material is covered in this recommended way.

Tackling the Examination Paper

1. Read the paper throughout before attempting to answer. (Some examination boards allow a preliminary period of 10 minutes or so for this exercise.)

2. Tackle first the question that appears to you as the easiest, rather than the most interesting or challenging.

3. If time allows, check through your answers before submitting the script.

4. If time is insufficient to complete an answer, a few words or figures indicating the result expected will help to show the examiner that you understand the question and its method of solution, and could probably answer it correctly if given time.

5. Remember that examiners invariably seek opportunities to merit a pass rather than a failure. Make their job as easy as possible.

Good Luck!

Appendix 2
Touch Typing

1. Touch typing is well worth learning because it is quicker than using two index fingers alone, and is less stressful to the eyes and neck.

2. Place the four fingers of the left hand on the keys ASDF, and the four fingers of the right hand on ;LKJ. These are the **Home Keys**, and the fingers should **always** return to them after any other letter. Reserve the right thumb to operate the space bar.

3. Type several times the letters ASDFGF (using the left first finger for G), and then ditto ;LKJHJ (using the right first finger for H), **returning the first fingers to their home keys of F and J each time.**

4. When proficient with this (and rhythm is vital), type R with the left first finger, **returning to its home key F.**

5. Similarly with V (returning to F), and then QWER and T, followed by ZXCV and B **returning to the home keys each time.**

6. Repeat this procedure with the right fingers, i.e. POIUY, followed by .,M and N, **returning to the home keys each time.**

7. To type upper case, use the Shift Key opposite the letter you are typing, e.g. if upper case D is required, use the right-hand Shift Key.

8. Practise, practise, practise, maintaining a steady rhythm (we learnt with military music as a background, and with covers over our keyboards). When you can type the alphabet smoothly without looking at the keyboard, you will have arrived.

Good typing.

Two words of warning:

1. The old style mechanical typewriters needed the fingers to give a certain amount of load to operate the keys. In the modern word processors, the keys are so hair-triggered that the mere weight of the finger can easily operate a key unintentionally.

   Hence, beware of unwelcome letters 'k' and 'd' as the middle fingers rest too heavily on their home keys, and adjust the tension of the keys accordingly where this facility exists.

2. When deleting any part of a line of text, the system can easily delete the previous line as well, for the same hair-trigger reason.

   Hence, check the highlighting before deleting with the space bar.

We were taught:

>Three spaces after a full stop;
>Two spaces after a semi-colon or colon;
>One space after a comma;
>Two spaces between words in upper case;
>Start the address halfway down the envelope.

Appendix 3
Matt Gregg's Method of Teaching Fuel Technology

Matt decided that the best way to identify the various fuels derived from petroleum was to arrange them in ascending order of density, ranging over typical values of 0.72 to 0.97 kg/L at 15°C from aviation gasoline through motor gasoline, kerosine, gas oil, and diesel fuel to residual fuel oils. He then treated each significant property in turn, plotting its value against density, resulting in a comprehensive figure showing the variations of all these properties with density. This indicated clearly how a change from one fuel to another would affect properties, and hence the problems that might arise.

As a personal interest, he compared the variations in flammability and ignitability of these fuels. He showed that the light gasoline-type (petrol) fuels are relatively flammable. Catching fire easily on the application of a flame (as determined in the flash point test) – which of course is common knowledge. On the other hand, the heavy residual fuel oils, although non-flammable, are relatively ignitable, catching fire in hot air without the application of any flame (as determined in the spontaneous ignition test) – which is not so widely recognised. He explained this by the fact that the larger fuel molecules rupture more readily through agitation in hot air, so exposing free bonds to react with atmospheric oxygen and promote ignition.

	Gasoline	RFO (Class F)
Flash Point, °C	-40*	100
Spontaneous Ignition Temperature, °C	470	300†

* Relatively flammable
† Relatively ignitable

All the test results referred to above are shown diagrammatically in Figure A3.1.

Typical data for petroleum fuels, alternative fuels and rocket oxidants are shown in Tables A3.1, A3.2 and A3.3 respectively.

Table A3.1 **Representative Petroleum Fuels**

Technical Name	General Name	Typical Density * kg/L @ 15°C	Main Applications
Aviation gasoline	Aviation petrol	0.72	Aero spark-ignition engines
Motor gasoline	Petrol	0.74	Automotive spark-ignition engines
Wide-cut fuel		0.78	Aero gas turbines for some military aircraft
Kerosine	Kerosene	0.80	Aero gas turbines, ramjet and rocket engines
High-flash fuel		0.82	Aero gas turbines for carrier-borne naval aircraft
Gas oil	Diesel	0.84	High-speed automotive compression-ignition engines, and industrial gas turbines
Diesel fuel	Diesel	0.86	Low-speed marine and power generating compression-ignition engines
Residual fuel oil	Fuel oil	0.97 to 1.01	Boilers and furnaces

* This terms replaces the earlier terms of 'Specific Gravity' and 'Relative Density' but the values are the same although no units were involved since they were ratios.

Figure A3.1 Composite plot of major properties of non-marine petroleum fuels

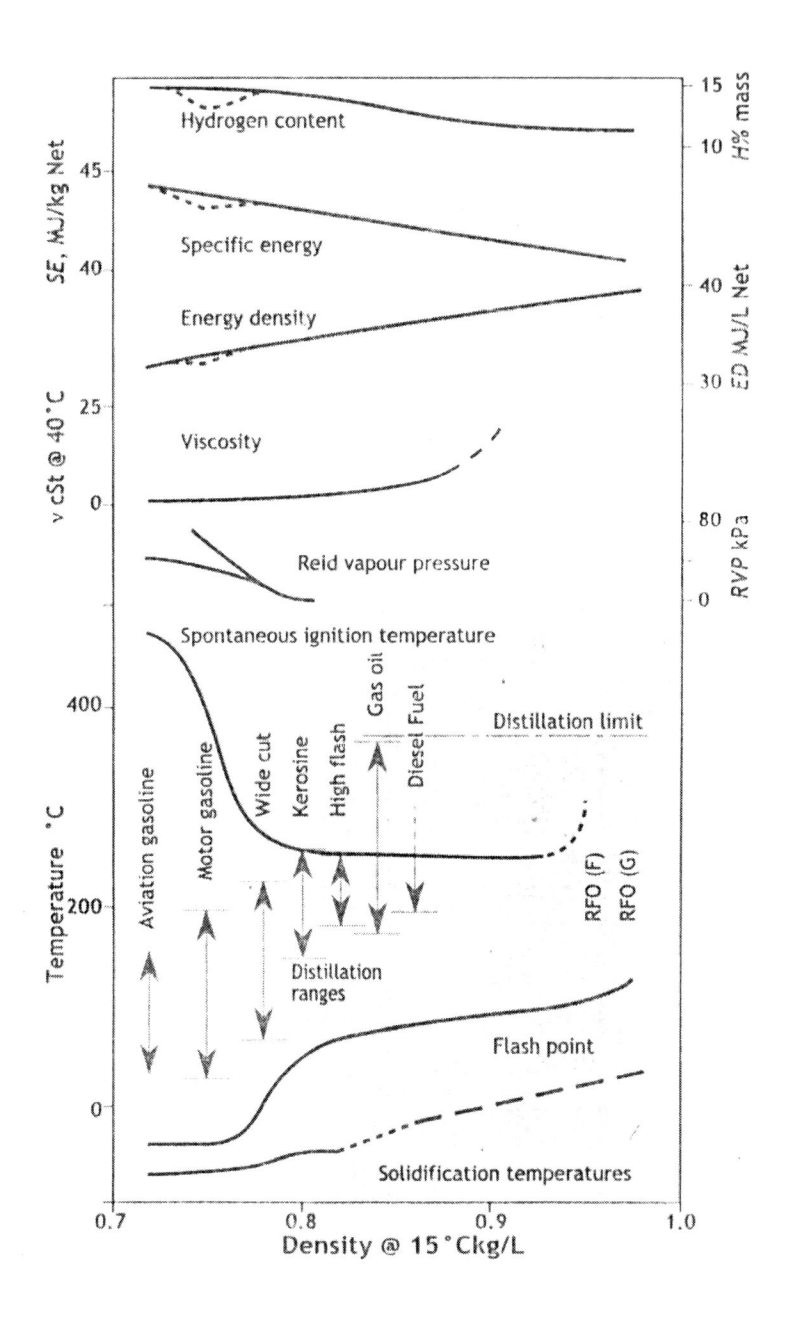

Table A3.2 **Representative alternative fuels**

Name	Formula	Main Applications
Compressed natural gas (CNG) (Methane)	$CH_4(g)$	Heavy duty lorries
Liquified petroleum gases (LPG) (Propane + Butane)	$C_3H_8 + C_4H_{10}$	Automotive spark-ignition engines
Methanol (Methyl alcohol, wood alcohol)	CH_3OH	Automotive spark-ignition engines
Ethanol (Ethyl alcohol, grain alcohol)	C_2H_5OH	Automotive spark-ignition engines
Hydrazine	N_2H_4	Rocket engines
Nitromethane	CH_3NO_2	Drag racing
Liquid hydrogen	$H_2(l)$	Fuel cells, scramjet* & rocket engines

* Supersonic combustion ramjet

Table A3.3 **Main Rocket Oxidants**

Name	Formula	Mass percent 'oxidant'
Liquid oxygen (LOX)	$O_2(l)$	100
Hydrogen peroxide (HTP)*	H_2O_2	94.1
Nitric acid (RFNA)†	HNO_3	76
Nitrogen tetroxide	N_2O_4	69.6
Liquid fluorine	$F_2(l)$	100
Chlorine trifluoride	ClF_3	100

* High test peroxide

† Red fuming nitric acid

For further information, refer to:

Transport Fuels Technology

E M Goodger, BSc(Eng)Hons, MSc(Eng), PhD,
CEng, CPEng, MIMechE, FEI, MRAeS, MIEAust

Landfall Press & Energy Institute (London), 2000

ISBN 0 952016 2 4

444 pages, 8 chapters, 58 appendices, over 200 figures (many coloured), 63 tables, bibliographies

Medium quarto, 235 x 276 mm   Current price £40

All enquiries to:

Energy Institute

61 New Cavendish Street, London,, W1G 7AR, UK
020 7467 7100
sfm@energyinst.org.uk
www.energyinst.org.uk

Appendix 4
The City of Bath (1988)

KEY TO HOTELS

A	ROYAL CRESCENT	E	LANSDOWN GROVE	I	DUKES
B	FRANCIS	F	ROYAL YORK	J	CARFAX
C	BEAUFORT	G	PRATTS	K	CHESTERFIELD
D	BATH	H	REDCAR	L	HARINGTON'S

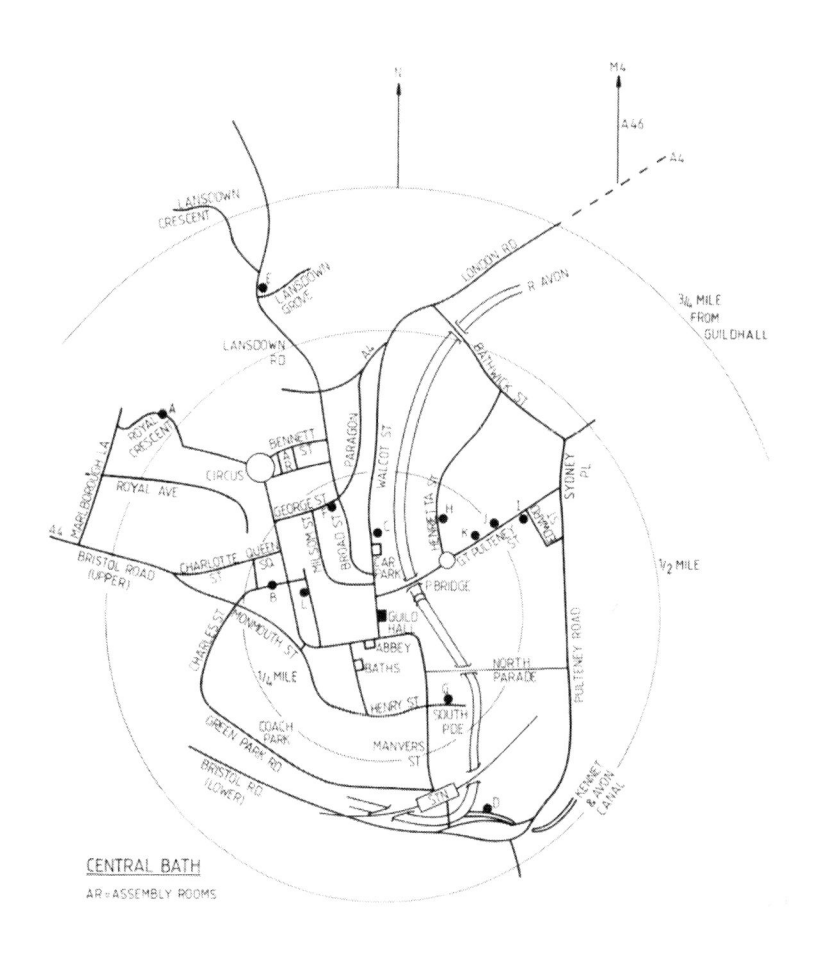

CENTRAL BATH
AR = ASSEMBLY ROOMS

Bath
Portrait of a City ---- in Pale Gold

"The prettiest city in the Kingdom" **Samuel Pepys**
"Oh, who can ever be tired of Bath" **Jane Austen**
"I like Bath: it has quality" **H V Morton**

Facts and Figures

Population:	85,000 approximately
Area:	7100 acres (2850 hectares)
Early closing:	Monday or Thursday, *but many shops open six days a week*
Road:	107 miles (172 km) west of London
	85 miles (134 km) west of Heathrow
	8 miles (13 km) south of junction 18 on M4 motorway
	14 miles (23 km) east south east of Bristol
Rail:	70 minutes from London, Paddington (hourly service)
Air:	Bristol Lulsgate Airport

To most of us, the name of Bath is synonymous with handsome Georgian architecture of sweeping aspect, remarkably well preserved plumbing of Roman vintage, a particular design of hand carriage for the aged or infirm, and such delicacies as Bath Buns, Sally Lunns, Bath Olivers and Bath Chaps. All this, and so much more, can be discovered by visitors even if their stay is brief.

The city of Bath rests in the deeply incised Avon valley between the Cotswold and Mendip Hills, firmly embracing its history within the ambience of its honey-coloured stonework. Many visitors arriving by road will approach southwards by the A46 from the M4 motorway and, soon after crossing the A420, will be treated on both sides to attractive vistas as the valley system unfolds. A gentle descent follows to the right-hand turn onto the London Road (A4), and thence past buildings on a towering cliffside towards the city centre. (The alternative, western descent through Lansdown, at 1/9 gradient is not so gentle but popular nevertheless.)

The first impressions likely to be gained are of elegant buildings of pale gold stone set in broad busy streets linked by picturesque precinct lanes with shopfronts rich in character. Hardly a brick is to be seen, the stone apparently being used for *all* buildings, not just those for municipal use as in other cities. This characteristic background colour stems from the ochre oolitic limestone quarried from the local downs and, although the walls of elderly cities invariably attract grime over the years, the cleansing activities in Bath (using water only to avoid damage!) are visibly effective, and are enhanced by the colourful splashes of flowers in every street.

From the multi-storey car park by the Beaufort Hotel in Walcot Street, one is led immediately on to the Guildhall, the Abbey Church and thus to the Roman Baths themselves, which appear to define a natural centre of a city that is notably compact, with no major point of interest more than about ten minutes walk from any other. The Florentine-style Pulteney Bridge, for example, is only three or four hundred yards to the north east, and even the famous Circus and Crescents are only minutes away, with much of the grandeur of the city lying within these landmarks. Those visitors arriving by rail will alight at Bath Spa station, and face a short walk up Manvers Street towards the Abbey. Scenery is invariably enhanced by reflections in water, and Bath is well served not only by the Avon itself, which frames it on three sides, but also by the Kennet & Avon Canal, the two joining forces at the foot of a flight of locks beside the Bath Hotel.

Early History

Any exploration of this geography will unfold a related pattern of history since no amount of bustle from present-day traffic and modern amenities can dull the impression of a living, city-wide museum. Bath's history is, in fact, founded on the presence of the hot springs which, from the discovery of flint implements, it is assumed were used by bands of hunters in about 5000 BC. According to legend, the first recognition of the significance of the springs occurred in 860 BC by Prince Bladud, son of King Hulibras and the father of King Lear. When this popular prince contracted leprosy he had to be banished from the court, and so became a swineherd. To his further horror, the pigs developed the same disease, but their instinct led them to wallow in the pools of warm mud, from which they could be enticed only by the sight of a bag of acorns. On finding the pigs cured of their skin ailment, Bladud shrewdly used the same medium to heal himself. Accepted back into court, and eventually gaining the throne, he founded a settlement on the site of this remedial spring as a token of his gratitude. (Interestingly, King Bladud emerged as a pioneer aviator, using a pair of home-made wings. Launching himself from a pinnacle he discovered, alas, that lift did not equal weight, and so met his end.)

Modern medicine is sceptical regarding curative powers of the spring waters, but the similarity of diseases suffered by men and pigs, and the trace of zinc in the waters that prevents some skin ailments, add strength to this legend. In any event, the Celtic tribe of the Dubunni subsequently held the waters of the hot wells as sacred to an ancient god, Sulis. After their invasion in AD 43, the Romans also recognised this therapeutic quality, and developed elaborate baths to create a spa known as Aqua Sulis (Waters of Sulis) which was used as a social centre as well as for bathing and curative purposes. They also merged their religion peacefully with that of the Celts by building a shrine dedicated to Sulis and also to Minerva, their goddess of healing. This was probably an act of reconciliation after putting down the bloody rebellion of Boudicca.

The Roman hydraulic engineers devised a very ingenious system of water control, with supply pipes and bath linings of lead mined from the Mendip Hills, and a bronze sluice normally held in the raised position but lowered periodically to flush away the black sand welling up with the water. Certain areas of the bath complex were heated artificially by the underfloor passage of hot air drawn from charcoal burners through vertical flues set in the walls. The centrepiece of the whole complex was the 80 by 40 ft Great Bath equipped with a barrel-vaulted roof made of hollow box tiles to reduce weight, and with open ends to permit escape of steam and so prevent cold droplets of condensed water falling onto the bathers below. After undressing and exercising, the bathers were oiled, sanded, stripped with a bronze strigil blade, cleaned and massaged, followed by immersion into the various baths in turn, all the time gossiping, being amused by entertainers, and attended by servants and slaves. However, the springs were also considered sacred, and Roman worshippers threw in votive offerings, including metal vessels, gem stones and a total of about twelve thousand coins. Interestingly, also found were about ninety examples of curses written on sheets of pewter, wishing impotence and death on those who had done the writer wrong. Should the wrongdoer not be known, then a list of possible suspects was given in the hope of the true culprit being included who would then live in fear of having been cursed!

The waters rising today into the baths complex probably fell as rain onto the nearby Mendip Hills some ten thousand years ago, seeping to a depth of about twelve thousand feet where it was warmed by earth heat up to 96°C before rising through Penny Quick fault to reach King's Spring at 46.5°C, and the Hot Bath at 48°C. Visitors will therefore find of interest not only the presence of some thirty different minerals in the spring waters, but also the quite substantial quantities of energy provided continuously by Nature for at least twenty thousand years, transported by water emerging at the rate of about one quarter of a million gallons each day. Although these waters are too rich in lime for direct use in central heating systems, they give an indication of the massive geothermal potential of hot rock schemes proposed for district heating and the like.

Following the departure of the Romans some 350 years later, the problems of silting and flooding could no longer be handled, and the spa succumbed steadily to settlement and decay. Five hundred years of the Dark Ages followed, increasingly dominated by Saxon migrants; however Christian monks under Saxon rule built new baths round the thermal springs, and also the great Abbey of St Peter which, in AD 973, was used by St Dunstan, the Archbishop of Canterbury, for the crowning of Edgar, the first King of all England. This coronation set the pattern for all subsequent ceremonies for crowning the kings and queens of England. The year 1088 saw the building on this site of a cathedral so immense that the present Abbey, created by Bishop King in 1503, stands over its nave. The cathedral spread into what is now known as the Orange Grove, so named to commemorate the visit by William of Orange in 1734. The black book of charges against the clergy compiled by Henry VIII's agent included reports of the monks leading immoral lives, and work stopped on the Abbey in 1539 during the Dissolution of the Monasteries. Some 480 tons of glass, iron and lead were then removed and sold to local merchants. The condition of the building became all too clear to Queen Elizabeth I when she and her godson, Sir John Harington, tried to shelter there during a rainstorm, and work restarted in 1574 after the Queen granted Royal Letters Patent ordering a nation-wide collection for repairs. The Abbey, an example of perpendicular English Gothic architecture, has a rectangular tower and impressive fan-vaulted interior. On the western end is a curious stone representation of Jacob's ladder featuring falling and rising angels, inspired by the Bishop's dream to restore the church. Its 52 windows have led to the description of the 'Lantern of England', two of the windows dating back to the 17th Century. In Georgian times, the bells were rung to announce the arrival of important visitors, who were then charged a fee of half-a-crown for the privilege, and invited to subscribe two guineas towards the social round of entertainments.

One of the oldest houses in Bath to escape the clearances of later years is the timber-framed Tudor building known today as Sally Lunn's House, named after the famous cook who arrived in 1680 and sold brioche buns baked to a secret recipe. The house is now a refreshment room and kitchen museum with its original faggot oven and baking equipment, the cellar showing remains of early historical periods. The buns are sold today following the 1966 rediscovery of the recipe in a panel above the fireplace.

Meanwhile, the popularity of the spa continued, with John Leland in the 16th Century describing the many bathers seeking relief from 'Lepre, Pokkes, Scabbes and great Aches'. Also, forthcoming was the direct advice 'If ye parts under ye midrife be grieved, sit up to ye navel, but if ye parts above the navel be diseased, sit in into the necke.' In the second Century, the Emperor Hadrian had already forbidden mixed bathing, but by the middle ages people of both sexes bathed together naked, by day and night ('Headquarters of Satan' – Charles Wesley) in full view of a rowdy public who were inclined to throw cats, dogs and even each other into the water on the slightest provocation. This skylarking led to legislation imposing fines of 3s.4d and 6s.8d respectively, for throwing a live beast or a clothed person. Henry VI had already complained about the practice of stealing bathers' clothing for ransom. However, Samuel Pepys (how appropriately named!) admired the 'very fine ladies' with whom he shared the bath.

Enter 'Beau' Nash

A major service to public health was rendered when the aforementioned Sir John Harington invented the water closet (hence the American term 'the John'). Prior to this, however, the lack of sanitation in Bath led to such comments as 'an unsavoury town', and 'the streets are dunghills'. Nevertheless, interest in the city as a spa was fostered in the 17th Century when visitors taking the cure included the consorts of both James I and James II, and a parallel interest stemmed from the aristocratic and rich who were bored with summers in London, and wished to find new venues for gaming. Queen Anne's visits in 1702 and 1703 attracted other influential people, and Bath became a centre for all those of quality and rank. However, the more wealthy of the spa patients were pursued by hordes of quack doctors, and the gamesters by cardsharps, pickpockets, duellists and other undesirables, the city itself becoming notoriously raffish. Avon Street, in particular, became the haunt of robbers, and the 'receptacle of fallen women'. Nevertheless, Lord Chesterfield, who had discovered the Bath waters to be no cure for deafness even when poured into the ear, preferred gambling with cardsharps who invariably paid their debts, rather than with gentlemen who didn't.

It was the prospect of making a killing at the gaming tables that induced an impecunious young Welshman, Richard Nash, to journey by coach and on foot in 1705 to this potential Shangri-La. Nash was bull-faced and of large, awkward build, earning him the backhanded compliment of 'the very agreeable oddness of your appearance', but he was generous and honourable despite his lifestyle, and had a certain dash that enabled him, for example, to accept a wager to ride naked through a village on the back of a cow. His undistinguished career by that time included study (terminated on the discovery of his secret engagement to a pretty local girl) at Jesus College, Oxford — where he still owes a bill; a commission in the guards, and a position as a law student in the Temple. Despite his success in organising a pageant to celebrate King William's accession, extravagance, boredom and hunger drove him towards a lucrative career as a gamester. Early winnings persuaded him to stay in Bath, and when Captain Webster, the gaming room Master of Ceremonies, died in a duel, Nash was the natural replacement.

'Beau' Nash quickly established orders of discipline, dress and behaviour, and so became the arbiter of taste for polite society, transforming the spa into a fashionable resort, with nightly performances of music, dancing and other social activities. Nash successfully mixed the nobility with the wealthy middle-class gentry, all of whom were intrigued by his outrageous but finely phrased insolence, and accepted his authority in such matters as banning swords, riding boots and ladies' aprons in the ballroom, and bringing the dancing to a close precisely at 11 p.m. each night so that the invalids could get some sleep. He prohibited duelling, but had to fight one himself beforehand as proof that this law was not for his personal protection. He persuaded the Corporation to improve the roads, and attempted to control the extortionate charges of the sedan chairmen. Known as the King of Bath, Nash sported a white hat as his crown, and travelled in a post chariot drawn by six greys with all the trappings of outriders, footmen and French horns. His many affairs with visiting ladies were offset to some extent by bursts of generosity to the unfortunate and unwise.

Nash took his cut from the gaming rooms, and flourished accordingly until the Gaming Acts of 1739 and 1745 forbade public gambling, and after being swindled out of much of his income, he was awarded a pension of a mere £10 a month by the Corporation. When he died in poverty in 1761 at the age of 88, he was awarded a splendid funeral and posthumous honours of every kind. His final mistress, Mrs Juliana Papjoy, spent the rest of her life gathering herbs and sleeping in a tree.

A Golden Age of Architecture

In parallel with the above events, a number of interesting archaeological finds were being made, including a remarkably well-preserved relief of Roman occupation in the shape of the gilded bronze head of Minerva (1727), and the pediment from the original Temple of Sulis Minerva in the form of a Celtic god, described by some as a Gorgon (1770). During this period also, the acceptable standards of English architecture were recorded in detail in a number of pattern books to which reference could be made by subsequent designers. Two famous names were associated with implementing the 'Rules of Taste' in the further development of the fabric of Bath. The first of these is Ralph Allen, a businessman who amassed a fortune by revolutionising the postal services of the nation. He also recognised the quality of the local stone that had been favoured by the Romans, and bought the quarry at Combe Down some two miles south of the city.

The second name is that of John Wood, a Yorkshire artist/architect devoted to the 16th Century Palladian style. On his arrival in 1727, he set about upgrading the medieval city by collaborating with Ralph Allen on a grand classical design in the style of Rome itself, both cities being surrounded by seven hills (known locally as 'downs'). The stone from Combe Down was selected for the purpose, despite criticism of its durability by London architects. Wood built Queen Square, and Ralph Allen's York Street house and Prior Park mansion. By the year of his death, he had started building the 315 ft diameter Circus with its three equal arcs of ten three-storied houses, each fronted by Doric, Ionic and Corinthian columns, and many motifs symbolising the arts, sciences and occupations. The central circle was originally paved with cobbles incorporating a grass-covered water reservoir, now replaced by splendid plane trees, and the stone acorns on the parapets commemorate the Bladud legend. The Circus was completed by his son, John Wood the Younger, who went on to create the Assembly Rooms in 1771, and in 1775 the thirty houses, fronted by 114 huge Ionic columns, comprising the majestic 600 ft semi-elliptical sweep of the Royal Crescent. Number One of this crescent was built for Thomas Brock, Wood's brother-in-law, and has now been given to the Bath Preservation Trust, serving as a museum.

Many other Georgian architects and designers have left their marks in the stonework of Bath. The Octagon was built in 1767 – but never consecrated as a chapel – and the year 1770 saw the completion of Pulteney Bridge, built by Robert Adam and named after the first Earl of Bath. This bridge over the Avon is lined with tiny shops on both parapets in the style of the Ponte Vecchio in Florence. Excellent views of the bridge, and of the three picturesque downstream weirs of parabolic appearance, can be gained from either bank, and from the North Parade Bridge. Since the fashionable Assembly Rooms were not accessible to wealthy tradesmen, they commissioned the building of the Guildhall in 1776 to the design of Thomas Baldwin, incorporating a magnificent Banqueting Room in Adam style, and still with its original crystal chandeliers and portraits of royalty by Sir Joshua Reynolds and William Hoare. Baldwin also designed the 1100 ft long Great Pulteney Street which, approaching 100 ft in width, is reputed to be the widest street in Europe. In 1789, John Palmer, the architect, built Lansdown Crescent, rivalling the majesty of the Royal Crescent. An earlier John Palmer, brewer and chandler, opened the Theatre Royal in 1750, and his son (also John) made a fortune by persuading the government to carry the mail via stage coaches rather than relays of mounted post boys.

The Pump Room, dating originally from 1706 as the spa's social and cultural centre, was rebuilt in 1796 as an elegant Corinthian-columned room, and is now a famous rendezvous to take a pleasant afternoon tea accompanied by music from the Pump Room Trio. The room houses the King's Spring from which spa water is still dispensed, and there is a statue of Beau Nash in a wall niche. Much of the furniture is original, including a Tompion clock presented by the maker to the city in 1709 together with a sundial to check its accuracy. In the spa's heyday, the daily dosage recommended was three glasses of spring water, and carved in Greek above the Pump Room's entrance is the motto "Water is Best". However, the descriptions by the indefatigable diarist, Celia Fiennes, as 'tastes like the water that boyles eggs' (together with her account of the daily removal of surface scum), and by Sam Weller of 'a wery strong flavour o' warm flat irons', are hardly supportive.

The opening of the Royal Mineral Water Hospital in 1742 by John Wood was supported financially by Beau Nash and Dr William Oliver, a leading physician who invented the Bath Oliver biscuit to offset the rich food. The £3 admission fee served to pay either the patient's fare home if cured, or for burial if not. Records show that 31% were cured, 46% were much improved, and only 3% died! This seems a remarkable achievement when it is recalled that only 150 years earlier the average life expectancy at Bath ranged from 55 years for Gentlemen and Professional Persons, to 25 years for Mechanics and Labourers (corresponding figures for Liverpool were 35 and 15!).

Frederic William Herschel brought fame to the city in a different way. This young German, after deserting from the military band of the Duke of Cumberland's army because of the harsh conditions, became conductor of the Assembly Rooms orchestra, and organist at the Octagon. His major interest, however, lay in astronomy, and using a telescope of his own manufacture, he discovered the planet Uranus in 1781. He was knighted and made Director of the Royal Astronomical Observatory, the King granting him a free pardon for his earlier desertion. Herschel's dedication to astronomy was shared by both his sister and his son, and his house is now a museum of his life and achievement.

The Last Two Centuries

The Victorian era saw the completion of the Kennet & Avon Canal in 1810 linking Bristol with London, and the mock-Tudor picturesque Bath Spa station – originally roofed – built by Isambard Kingdom Brunel in 1841 for his Great Western Railway, but neither enterprise was outstandingly successful in improving prosperity. Certain industries and inventions were centred in the city, as indicated in the chronological chart. The famous indoor market was built in the Guildhall during this period. Sir Isaac Pitman published his 'Stenographic Sound Hand' in 1844. Then in 1878 the Great Bath was discovered during an investigation of a leak from the King's Bath, the site being excavated and embellished with masonry and statues. A unique contribution to Bath is the Empire Hotel built in 1901 by the City Surveyor, Major Davis who, having been cheated of his chance to roof the Great Bath, appears to have seized the opportunity for a last monument to Victorian Bath in this edifice with a skyline featuring a castle, a manorial house, and a cottage so that guests of all classes could feel welcome. Nowadays, the Empire Hotel is occupied by the Admiralty.

The hundreds of memorial tablets to be found in the Abbey, together with a large number of plaques on house fronts, bear witness to the many famous residents and visitors. In addition to those mentioned above, these include: Jane Austen, Edmund Burke, Fanny Burney, Lord Clive, Charles Dickens, Benjamin Disraeli, John Evelyn, Henry Fielding, William Friese-Greene (of cinematographic fame), David Garrick, Thomas Gainsborough, Oliver Goldsmith, Lady Hamilton, Sir Henry Irving, Dr David Livingstone, Franz Liszt, Lord Macaulay, the Emperor Napoleon III, Lord Nelson, Lord Palmerston, Admiral Phillips (Governor of New South Wales), William Pitt, Sir Walter Scott, Richard Brinsley Sheridan (who scandalised society by eloping with 17-year-old Elizabeth Linley, marrying her the following year after surviving two duels), Sarah Siddons (owner of the Theatre Royal), Queen Victoria, John Wesley (who preached at Huntingdon Centre), William Wilberforce, William Wordsworth and, latterly, Sir John Betjeman. One imagines the countless legions of artists, writers, irascible colonels and maiden aunts who have all contributed to the 'busy idleness' of Bath.

The twentieth century added its own chapters to the story of the city. A choir unit was added to the Abbey in memory of World War I. In 1925, the Bath Act ruled that a substitute for the over-expensive Bath stone must be used for facing new buildings, and crushed limestone was adopted as a basis. World War II brought three bombing raids within two nights in 1942, killing over 400 people, damaging 3000 buildings, and gutting the Assembly Rooms which had been refurbished and given to the National Trust only a few years earlier. Although the Assembly Rooms have been meticulously restored, much of the remaining redevelopment has led to criticism. Fortunately, a more rational approach to rebuilding appeared in 1978 with the publication of 'Saving Bath'. The Alkmar Gardens in the Orange Grove commemorate the liberation of the Netherlands in 1945, and an annual music festival was founded in 1947. The Bath Excavation Committee, later called the Bath

A CHRONOLOGY OF BATH

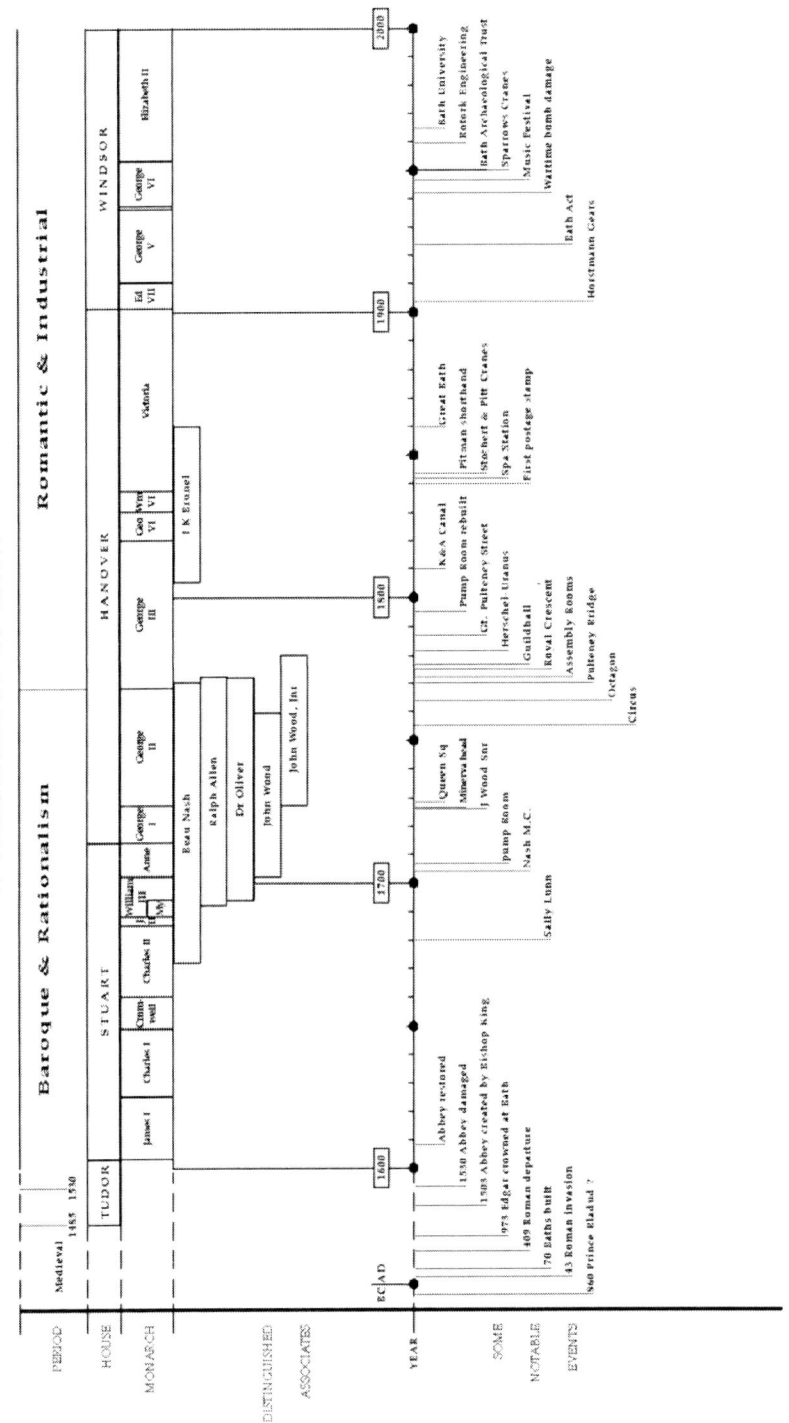

Baroque & Rationalism | Romantic & Industrial

PERIOD: Medieval 1455-1530

HOUSE: TUDOR | STUART | HANOVER | WINDSOR

MONARCHS:
- James I
- Charles I
- Cromwell
- Charles II
- James II / William III / Mary
- Anne
- George I
- George II
- George III
- Geo IV / Wm IV
- Victoria
- Ed VII
- George V
- George VI
- Elizabeth II

DISTINGUISHED ASSOCIATES:
- Beau Nash
- Ralph Allen
- Dr Oliver
- John Wood
- John Wood, Jnr
- I K Brunel

YEAR: BC/AD — 1600 — 1700 — 1800 — 1900 — 2000

SOME NOTABLE EVENTS:
- 500 Prince Bladud ?
- 43 Roman Invasion
- 70 Baths built
- 409 Roman departure
- 973 Edgar crowned at Bath
- 1088 Abbey created by Bishop king
- 1150 Abbey damaged
- Abbey restored
- Sally Lunn
- Pump Room
- Nash M.C.
- Queen Sq
- Minerva head
- J Wood Snr
- Circus
- Octagon
- Pulteney Bridge
- Assembly Rooms
- Royal Crescent
- Herschel Uranus
- Guildhall
- Gt. Pulteney Street
- Pump Room rebuilt
- K&A Canal
- Great Bath
- Pitman shorthand
- Stothert & Pitt Cranes
- Spa Station
- First postage stamp
- Horstmann Gears
- Bath Act
- Wartime bomb damage
- Music Festival
- Sparrow's Cranes
- Bath Archaeological Trust
- Rotork Engineering
- Bath University

367

Archaeological Trust, was established in 1950, and the technologically-oriented Bath University was created at Claverton Down in 1965. Happily, the 1978 problem of amoeba contamination of the spa waters has been solved by means of a deeper borehole.

Most of the architectural and cultural treasures of Bath can be discovered by independent exploration, preferably armed with an information booklet from the City Council. However, limited time can be used optimally by joining either a free guided tour offered by the Mayor's Corps of Honorary Guides, or by venturing slightly further afield with a guided coach tour.

Bibliography

The following acknowledgements are made gratefully:

Chapter 5.

The Bombed Fuel Tank Story, Shell Petroleum Co. Ltd, Shell Haven.

Chapters 5 onwards. Reminders of World Events

Anon. *125 Years in Words and Pictures*, The Daily Telegraph, 1855-1980.
Penny Vincenzi, *Taking Stock*, Collins Willow Books, London 1985.
Edward Townley, *The Complete A-Z 20th Century European History Handbook*, Hodder & Stoughton, 2000, ISBN 0 340 67996 4
Michael Oke, *Times of Our Lives*, How To Books, Oxford, 2005.

Chapter 9. Reminder of Operation Overlord.

Anon, *Invasion Without Tears* leaflet, 84 Group Rear Headquarters, 2nd TAF.

Chapters 9 & 19. Reminder of PLUTO.

Arthur Clifford Hartley, *The Engineer at War* – a symposium of papers on war-time engineering problems, Institution of Civil Engineers, London, 1948.
Adrian Searle, *PLUTO, Pipe-line under the Ocean*, Shanklin Chine, IOW, 1995, ISBN 0 9525876 0 2

Chapter 10. Reminder of V2 details.
John Humphries, *Rockets and Guided Missiles*, Ernest Benn Ltd, London, 1956.

Chapter 13. Grand Union Canal.

Hugh McKnight, *The Shell Book of Inland Waterways*, David & Charles, London, 1975, ISBN 0 7153 6884 2

PJG Ransome, *The Archaeology of Canals*, World's Work Ltd, Tadworth, 1979, ISBN 437 144003

Chapters 13, 16, & 19 Histories of selected towns

Anon, *The Illustrated Road Book of England & Wales*, The Automobile Association, London, 1965

Anon, *AA Illustrated Guide to Britain*, Drive Publications Ltd, Basingstoke, 1972.

Anon, *AA Book of British Towns*, Drive Publications Ltd, Basingstoke, 1979.

Countless conversations with the late William Woodrow, native of Norwich

Michael Gibbs, *Norwich*, Harrap, London, 1991, ISBN 0 245-60256-9

Christopher Smith, *Norfolk and Norwich Millennium Library*, The Forum, Millennium Plain, Norwich, NR2 1AW

Chapter 16. French terms.

Ron Wingrove and John Jammes, *Aviators' English and French Dictionary*, Cranfield Press, 1985

Chapter 18. History of Bath.

'Panacea: or the Universal Medicine', *QJM – An International Journal of Medicine*, July 2005. Front paper.

Reginald W M Wright, *Bath Abbey*, Pitkin Pictorials, London, 1973, ISBN 0 85372 054 1.

Diana Winsor, *The Dreams of Bath*, Trade & Travel Publications Ltd, Bath, 1980, ISBN 0 900751 16 9

Anon, *The New Bath Guide*, Ashgrove Press Ltd, Bath, 1985, ISBN 0 906798 44 2

Anon, *The Roman Baths & Museum*, Bath Archaeological Trust, Bath, 1985, ISBN 0 9506180 12

Tricia Simmons & Marion Carter, *Bath*, Unichrome, 1986, ISBN 0 9507291 08

Barry Cunliffe, *The City of Bath*, Alan Sutton Publishing Ltd, Gloucester, 1986, ISBN 0 86299 250 8

Paul Newman, *Bath*, The Pevensey Press, Cambridge, 1986, ISBN 0 907115 32 2

Edith Sitwell, *Bath*, Century, 1987, ISBN 0 71261507 5

Chapter 19.
Newhaven/Eastbourne story: Professor F J Bayley, Brighton.
Cockerel story. Unknown customer of a Norfolk pub.

All Chapters. General advice on writing life stories.

Nicholas Corder, *Writing Your Own Life Story*, Straightforward Publishing, Brighton, 2004, ISBN 1903909 48 1

Michael Oke, *Times of Our Lives*, How To Books, Oxford, 2005.

All Chapters. Information on developments in fuels technology

E M Goodger, *Transport Fuels Technology*, Landfall Press & Energy Institute, London, 2000, ISBN 0 9520186 2 4

E M Goodger, TFT Update Service, Ongoing, The Energy Institute, 61 New Cavendish Street, London, W1G 7AR

Index

Airbus A380 aircraft	333
Airscrew	55
Alcohol fuels	119, 218, 225, 226
Alternative fuels	218, 221, 224, 226, 227, 326, 348, 350
Ammonia fuel	218
Apollo	218, 219
Armstong Whitworth Whitley aircraft	42
A-symmeTree	144
ATA (Air Transport Auxiliary)	40
Aviation fuel	243
Avro Anson aircraft	266
Bath	243, 244, 353-368
Battle of Britain	15, 111
Beaufighter aircraft	64
Biggin Hill	133, 144
Biofuel	226, 227
Black market	313
Bletchley	32, 62, 63, 66, 238, 337
Boeing 737 aircraft	266
Boeing 777 aircraft	333
Bromine	52
Canals	153-168
Canberra (ship)	194, 200, 230
Carburettor	48, 75, 147
Cat. AC	46
Cat. E	23, 46
CFR (Co-operative Fuel Research)	50, 123
Challenger	235
Channel Tunnel	267
Chocolate biscuit	235
Churchill, Sir Winston	113, 122, 123, 144, 146, 195, 230, 271
Clean air policy	135
Clipped wing Spitfire aircraft	96
Coal	85, 135, 190, 218, 226, 235, 359

Coal-derived liquids	218
Columbia	230
Comet (1)	152, 167
Comet (2)	323
Concorde	194, 210, 214, 222, 224, 225, 315
Concordski	222
Crocodiles	240
Derwent gas turbine	114
Detonation	50
Diesel fuel	150, 151, 157, 163, 194, 226, 239, 240
	242, 254, 276, 310, 311, 335, 347, 248
Diesel knock	221
Doodlebug	101
Dopes	51
Droplet rig	236
Eland gas turbine	114
Energy Institute	310, 351
Engineering	38, 43, 44, 45, 46, 53, 74, 78, 83, 119, 131,
	163, 184, 198, 206, 215, 231, 232, 271, 289,
	291, 293, 294, 303, 308, 309, 310
Engine handling	43, 47, 57, 63, 73
Entropy	231, 291, 292
Ethanol (Ethyl alcohol)	119, 350
Ether	18, 218
Ethylene dibromide	52
Ethylene glycol	59
EVT (Educational & Vocational Training)	117-130
Farnborough	149, 207
Festival of Britain	144
Flammability	149, 152, 347
Fletcher trolley	131
Forum, The	276
Forward Repair Unit	84
Fuel cell	326, 350
Fuel injection	148
Fuel technology	163, 216, 239, 267, 270, 291, 301, 326, 347-352

Gasoline	110, 111, 114, 115, 347, 348, 349
Geothermal energy	291, 359
Glycol leak	59
Grand Union Canal	155, 238
Griffon aero engine	108, 110
Grosvenor House aircraft	323
HAIS cable	277
HAMEL steel pipe	278
Handley Page aircraft	133
Hards, The	102
Harnser	264
Harrier aircraft	207
Hawker aircraft	55, 133
Heptane, normal	50
High boost - low revs	47, 75, 138
Hurricane aircraft	39, 64, 144
HTP (hydrogen peroxide)	175, 176, 350
Hybrid cars	326, 335
Hydrazine hydrate	176, 350
Hydrogen	119, 218, 226, 350
Hypergolic	176
Ignitability	152, 347
Imperial Airways	323
Inheritance Tax	316, 319
Inland Revenue	317
Institution of Engineers Australia	310
Institution of Mechanical Engineers (i.e. 'Mechanicals')	310
Iodine	51
Jet aircraft	114, 152, 225
Jumbo jet aircraft	219, 225
Jungle juice	151, 172
Juno Beach	103
Kerosine	51, 114, 115, 225, 347, 348
Knock, spark	50, 218, 221, 311
[3], diesel	221
Knockmeter	123

Landfall	185, 205
Land mines	104, 131
LCVP (Landing Craft, Vehicle-Personnel)	102, 103
Lead oxide	52
Legging	163
Life Index	226
Liquefied gaseous fuel	218, 226
Liquid oxygen	119, 174, 175, 350
LNG (liquefied natural gas)	226
LPG (liquefied petroleum gases)	226, 350
Liquid hydrogen	119, 350
Luftwaffe	104, 115, 212
MAC (Merchant Aircraft Carrier) ships	170
Maldistribution project	147, 150
Merlin aero engine	44, 47, 49, 108, 110, 147, 171
Messerschmidt 109 aircraft	15
Meteor aircraft	114
Methanol (methyl alcohol)	110, 237, 350
Methyl aniline	110
Mine fields	104, 112, 131
Misfuelling	311
Mixture distribution	146
Moon walk	214
Motor gasoline (petrol)	110, 347, 348
MT (Motor Transport)	61, 84, 99, 101, 102
Mustang aircraft	44
Natural gas	226, 350
Nitric acid	174, 350, 351
North Sea oil and gas	224
North Sea pipe-line	224
Nosey Parker	262
Octane, iso	28, 50, 51, 61, 72, 276
Octane number	49, 50
OTU (Operational Training Unit)	34, 37-44, 46, 71, 83, 84, 93
Overlord	99-116, 293, 313, 369

Performance number	49, 50
Permanganate	119, 175
Petrol (gasoline)	49, 50, 51, 52, 61, 64, 67, 72, 73, 104, 110 173, 237, 239, 277, 311, 347, 348, 349, 350
Pioneer 10	231
Piper Alpha	243
PLUTO (Pipe-line under the ocean)	104, 277, 369
Propeller	40, 48, 49, 55, 75, 93, 103, 108, 109, 110, 114, 159, 205, 240, 241, 255, 264
Pull's ferry	264
Pulse jet	110
Queen Mary long low loader	29
R101	323
Rapeseed oil	226
Re-entrée	7, 144, 237, 246, 327
Rolls Royce	44
Rotary engines	93
Royal Aeronautical Society	310
Royal Leamington Spa	160, 165
RSU (Repair & Salvage Unit)	84, 91, 92, 93, 99, 100, 101, 110, 114
Sausage machine	102
Scavenger additive	52
Science Museum	139
Selenium	51
Shadow scheme	53
Single lever control	76
Smart mines	104
Smog	134
Smoke-free zones	135
Spaceflight	177, 194, 207, 215, 217
Spark knock	50, 218, 221, 311
Spitfire aircraft	15, 16, 29, 39, 40, 41, 65, 72, 108, 144
Spontaneous ignition	52, 178, 218, 236, 347
Sputnik	173
Strine language	89, 187
Supercharger	48, 49, 75, 148, 171
Surface Ignition	171

2nd TAF (Tactical Air Force)	84, 369
Technology	9, 119, 131, 163, 216, 231, 239, 267, 270
	277, 291, 301, 309, 315, 326, 347-353, 371
Tellurium	51
Tempest aircraft	110
TEL (Tetra-ethyl lead)	50, 51, 52
Thermodynamics	131, 148, 216, 301
Tiger Moth aircraft	41, 72
Trident aircraft	219
Twin jet apparatus	176
Typhoon aircraft	110
U-boat	109, 170
Unleaded fuel	239
V1 flying bomb (doodlebug)	101, 110, 114, 115, 123
V2 rocket	7, 113, 115, 118, 369
VE Day	115
Vegetable oils	218
VJ Day	118
Warwick	160, 164
Whitley aircraft	41, 42, 96
Wide-cut fuel	115, 348
Wing	15, 16, 41, 64, 65, 84, 91, 96, 100, 101, 103, 209, 358